APPLIED PHOTOGRAPHY
2nd Edition

ERVIN A. DENNIS

Delmar Publishers Inc.

NOTICE TO THE READER

Cover design by John Desieno
Cover photo by Randall Perry

New Products Acquisitions: Mark W. Huth
Developmental Editor: Sandy Clark Gnirrep
Project Editor: Andrea Edwards Myers
Production Coordinator: Bruce Sherwin
Art Coordinator: Michael Nelson
Design Supervisor: Susan C. Mathews

For information, address Delmar Publishers Inc.
2 Computer Drive West, Box 15-015
Albany, New York 12212

Printed in the United States of America
Published simultaneously in Canada
by Nelson Canada,
A division of The Thomson Corporation

10 9 8 7 6 5 4 3 2 1

Library of Congress Cataloging in Publication Data

Dennis, Ervin A.
    Applied photography/Ervin A. Dennis.—2nd ed.
        p.   cm.
    Includes index.
    ISBN 0827349114
    1. Applied photography.   I. Title.
TR624.D46   1993
770—dc20
                                                        91-33017
                                                        CIP

CONTENTS

PREFACE

Photography is one of the most effective communication forms known to the human race. There are no language barriers among the many countries, nationalities, and ethnic groups when photographic prints and transparencies are used to communicate. Cameras, film, chemistry, and people combine to deliver the news in pictures. Nearly everyone can see and learn something by looking at a photograph.

This 2nd edition of *Applied Photography* is designed to meet the needs of people interested in teaching and learning about continuous tone photography. This text clearly illustrates the relationships of photography and chemistry, electronics, computers, and technology. Creativity and photographic composition are also stressed. The technical and creative content of photography is presented in detail in the nine chapters and forty-four units of this book. *Applied Photography* is profusely illustrated with over 550 photographs and line drawings that show photographic results and the procedures used to achieve acceptable results. Learning is enhanced by using visual examples that portray both acceptable and unacceptable results.

Objectives, key terms, and study questions help to make each unit a self-contained study of a topic in photography. The content is organized logically to take the student photographer through a series of discussions and experiences to form a solid base of knowledge that will give the confidence needed for a life-long use of photography. For example, Chapter 1 contains five units designed to introduce students to the vast subject of photography. How photography is used in society and career opportunities are highlighted. Chapter 2 consists of six units that describe basic cameras through the several camera categories. This is a technically oriented chapter.

Chapters 3 and 4 contain a series of twelve units with descriptions on how to take pictures and light scenes. Cameras don't take pictures unless they are directed and controlled by humans. Black-and-white film and print processing are thoroughly covered in Chapters 5 and 6. Equipment, tools, chemistry, and procedures are strongly emphasized in the twelve units.

Color photography is highlighted throughout the four units of Chapter 7. Full color examples, theory, and step-by-step procedures provide enough information for a good foundation in producing color prints and slides. Chapter 8 includes content which stresses finishing, displaying, and preserving photographs. Finally, the units of Chapter 9 are utilized to introduce instant and electronic photography. Instant photography is in common usage and electronic photography is developing rapidly.

Photography is an exciting medium of visual communication. I wish you every success while using this text to obtain an understanding of the universal language—photography.

Ervin A. Dennis

ABOUT THE AUTHOR

Dr. Ervin A. Dennis is a professor in the Department of Industrial Technology, University of Northern Iowa, Cedar Falls, IA. 50614-0178. He has had a long and varied career in education, including four years of high school teaching and thirty years of teaching at the university level in four different institutions. The author has served four, three-year terms on the Cedar Falls, Iowa Board of Education. Dr. Dennis has authored several other publications including two major graphic arts textbooks and two overhead transparency series in graphic arts and photography. He has served in leadership rolls in several professional associations including the International Technology Education Association, Council on Technology Teacher Education, International Graphic Arts Education Association, National Association of Industrial and Technical Teacher Educators, Iowa Industrial Education Association, and Waterloo Club of Printing House Craftsmen. Currently, he is chair of the Policy Committee of the Technology Education Division, American Vocational Association.

ACKNOWLEDGMENTS

The 2nd edition of *Applied Photography* is dedicated to my wife, B. LaVada Dennis, who has provided much support throughout the many months while this rewrite was being prepared. Her untold number of hours of typing, keyboarding, filing, and researching made it possible for her author husband to complete this revision. To our three sons, Barton, Aaron, and Seth, I express appreciation for their support through photographic contributions and moral support. Never could I have completed a textbook, student manual and instructor's guide without their untiring encouragement and direct assistance.

Many individuals provided valuable assistance during the preparation of the 1st edition. These people included Mr. Thomas E. Sampson who served as a volunteer reader critic; Mr. Richard J. Kramer, Sr. who permitted the author to utilize several of his photography students at the Hawkeye Institute of Technology; and Mr. Sammie Dell of the Hawkeye Institute of Technology for coordinating the models utilized in many photographs.

Special recognition is given to five students who have now graduated from the Hawkeye Institute of Technology, Waterloo, Iowa for their efforts in supplying photographs. Their sincerity, dedication, and expertise are appreciated. The five student photographers were Miss Marla Dory, Mr. Tim Eastman, Mr. Alan Jackson, Mr. Brook Lightner, and Mr. Mark Reed.

For the preparation of this edition, recognition is given to the editors and production personnel at Delmar Publishers Inc. for their assistance, encouragement, and support throughout the many months of planning, researching, and writing required to produce this textbook. My student secretary, Miss Stacie L. Oetken, deserves special recognition for her assistance in readying each page of the three publications for revision work. Her dedication to task is greatly appreciated.

My photography students at the University of Northern Iowa have contributed much to my understanding of photography. They have given me the opportunity to design and improve educational materials. Most of all, they have given me a reason to be a "student" of photography. A special thanks to each and every former student.

The following reviewers evaluated the entire manuscript:

Charles H. Goodwin
Union Endicott High School
Endicott, New York

C. Michael Harmon
Tolono, Illinois

Orville Emmette Jackson
Sam Houston State University
Huntsville, Texas

Alan Towler
Department of Technology
Southwest Texas State University
San Marcos, TX

George R. Roland
Lincoln High School
Dallas, TX

Gordon C. Gunlock
St. Louis Park Senior High School
St. Louis Park, MN

L. Grant Luton
Schrup Middle School
Akron, Ohio

H. Warren King
Reseda High School
Reseda, CA

Lawrence Shapiro
Venice High School
Los Angeles, CA

James W. Bray
Cross Keys High School
Atlanta, GA

Roy Bartlett, Jr.
Austin High School
Austin, TX

Sandra Rodgers
Lowndes County High School
Valdosta, GA

Geoffrey Chandler
Franklin Central High School
Indianapolis, Indiana

Charles Foreman
South Carroll High School
Sykesville, MD

Robert K. Hartzell
Boyertown Area Senior High School
Boyertown, PA

David Richardson
Petaluma High School
Petaluma, CA

Bruce J. Mackie
El Camino Real High School
Woodland Hills, CA

Ken Maurizi
Galesburg High School
Galesburg, IL

The following photography instructors reviewed the table of contents and several selected units from the manuscript and offered their suggestions for content format and design:

Michael Passaic
Newburgh Free Academy
Newburgh, NY

Allen Saathoff
Sydney Lanier High School
San Antonio, TX

One hundred fourteen companies, institutions, and individuals provided photographs and line drawings for this publication. To all, I express sincere gratitude.

AGFA Corporation
American Airlines
Armed Forces Institute of Pathology
Bel-Art Products
Bencher, Inc.
Berg Color-Tone, Inc.
Berkey Marketing Companies
(Charles) Beseler Company
Bestwell Optical Instrument Corporation
Byers Photo Equipment Company

California Stainless Mfg.
Calumet Photographic, Inc.
Canon U.S.A. Inc.
Celestron International
Chinon America, Inc.
Clark's Photo Art Studio
Colenta America Corporation
David W. Coulter
Creatron Inc.
Barton A. Dennis
E. A. Dennis
(The) Denny Mfg. Co. Inc.
Dial-A-Photo System
Dimco Gray Company
Doran Enterprises, Inc.
Marla Dory
Eastman Kodak Company
Edric Imports, Inc.
Electronic Photography and Publishing, Sony
    Business and Professional Group
ESECO Speedmaster
Anthony F. Esposito, Jr.
Sharon K. Fahey
Falcon Safety Products, Inc.
GMI Photographic
René C. Gallet
Julie Habel
Hadland Photonics, Ltd.
(Victor) Hasselblad
Heico Chemicals, Inc.
(Karl) Heitz, Inc.
(The) Holson Company
Hope Industries, Inc.
Ilford, Inc.
Instrumentation Marketing Corporation
Alan Jackson
Jobo Fototechnic, Inc.
Richard L. Johnson
JVC Company of America
J. R. Karsnitz, *Graphic Arts Technology*
Birdie Kramer
Kelvin K. Kramer
Kinetronics Corporation
Kleer Vu Plastics Corporation

Kreonite, Inc.
Labex Engineering Corporation
Leedal Inc.
Leica Camera, Inc.
(E.) Leitz, Inc.
Light Impressions Corporation
Lisco Products Company
Louisiana State University
    Medical Center, Department of Pathology
Lowel-Light Mfg. Inc.
LTM Corporation of America
3M
Mamiya America Corporation
Michael Business Machines, Corp.
Minolta Corporation
(The) Morris Company
MWB Industries, Inc.
Mystic Color Lab
National Aeronautics and Space
    Administration
Nikon, Inc.
Noritsu America Corporation
Novatron of Dallas, Inc.
Olympus Corporation
Omega Arkay
Optische Werke G. Rodenstock
Pentax Corporation
Photo Control Corporation
(The) Pinhole Camera Company
Polaroid Corporation
Pete J. Porro, Jr.
Porter's Camera Store, Inc.
Print File, Inc.
Quantum Instruments, Inc.
Queen City Plastics
Richcolor Systems
James A. Riggs
Santa Fe Railway
(The) Saunders Group
Richard Schneck
Schneider Corporation of America
Marie Schreffler
Seal Products, Inc.
Siegelite Flash Bracket Co.

Sima Products Corporation
Rulon E. Simmons
Sinar Bron, Inc.
Randy L. Slick
Sports & Photo
Tamron Industries, Inc.
Tenba, Inc.
Charles C. Thomas, Publishers
Tiffen Manufacturing Corp.
Time-Field Co. (David M. Pugh)

University Products, Inc.
U.S. Department of the Navy
Vivitar Corporation
West Virginia University,
    School of Journalism
John Wolff
WW, Inc.
Yashica, Inc.
(Carl) Zeiss, Inc.

CHAPTER ONE

PERSPECTIVE OF PHOTOGRAPHY

WHAT IS PHOTOGRAPHY?

OBJECTIVES   *Upon completion of this unit, you will be able to:*
- *Discuss photography as a communicator.*
- *Explain why photography is a universal language.*
- *Cite examples of where photography has been used to record history.*
- *Describe how photography meets the recreational needs of people.*

KEY TERMS   *The following new terms will be defined in this unit:*
Photography
Universal Language
Visual Communication

INTRODUCTION

Photography plays an important role in the instant information age of today. Photographic images are used to record everyday events. Once prints are made, the results can be sent nearly anywhere in the world through electronic means.

THE WORD

Photography means "to write or draw with light." Without light, there can be no photography. It is necessary to precisely control light to achieve quality photographic results. There are other important aspects of photography, Figure 1-1. Each has its place in the total process of producing photographic images. All ingredients of the "Photography Transmission Chain" are presented throughout the following pages of this book.

COMMUNICATOR

For centuries, there has been a strong need for people to communicate. This need has increased over the years. Today, as never before, people rely on other people, equipment, and systems to provide up-to-the-second audio and visual information.

Photography has a central place in all communication, Figure 1-2. Educators and students benefit by using photographic slides to teach and learn. No other medium can record images of higher quality either in black and white or full color.

Photographs record much and tell much. There is considerable evidence of this fact as reported in family photo albums, Figure 1-3. Often when pictures are taken, there seems to be little meaning. After several years pass, they have considerable meaning. The recorded visual images provide a communication link that cannot be compared to other forms of communications.

UNIVERSAL LANGUAGE

Pictures communicate with people throughout the world, Figure 1-4. Photographs taken in an English-speaking environment can be understood by people who do not know the language. The reverse is also true. A picture taken in Germany or Japan is easily understood in the United States. Adding a few words in the native language gives specific meaning to any photograph.

Companies use photographs to advertise their products, Figure 1-5. The visual image captured on light-sensitive film and paper provides an exacting picture of the product. This is second to having the real product in hand. A photograph nearly permits the viewer to reach out and touch the product.

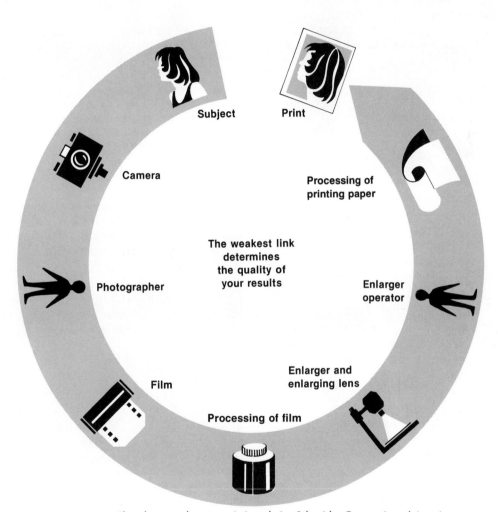

FIGURE 1-1. The photography transmission chain. *Schneider Corporation of America*

FIGURE 1-2. Pictures communicate information for students to learn and educators to teach. *Eastman Kodak Company*

FIGURE 1-4. This photograph serves to effectively communicate a common form of transportation available at this large resort hotel. *WW. Inc.*

FIGURE 1-3. Family photographs record the feelings and beliefs of individual family members.

FIGURE 1-5. A product is made almost "real" to the viewer through a photograph. *René C. Gallet*

RECORDER OF HISTORY

Photography helps to tell the facts of yesterday, Figure 1-6. The photographer with a camera in hand at the right time and place can record events that will live on forever. Words have and continue to be used to describe historical events. Unfortunately, even the best use of words cannot equal the detail and clarity of the photographic image.

Photography has a significant history in itself, but it has and continues to be used as a recorder of history. Two pioneers in the development and use of photography were George Eastman and

FIGURE 1-6. Photographers helped to record the many events of the United States Civil War during the early 1860s. *Armed Forces Institute of Pathology*

FIGURE 1-7. George Eastman (*left*) and Thomas Edison in the garden of Eastman's home during the 1928 announcement of the first Kodak color movie film. Kodak's 1899 development of long rolls of film that measured up to 1,000 feet (305 m) aided Edison in his development of a practical system of motion pictures. *Eastman Kodak Company*

The cave dwellers of thousands of years ago knew the value of visual images. Their drawings on cave walls served to tell people who followed them what they saw and did. These methods were crude, but their purpose was the heartfelt need to communicate by visual means.

Centuries later, people began to have the opportunity to record their own history in an easier way, Figure 1-8. They could take a picture on light-sensitive film in a matter of seconds. The recorded image provided detail and clarity that allowed the event to live on forever. No longer did happenings live only in the minds of people who were witnesses; happenings could be shared with people near and far. Also, for the first time, there was a way to share great detail and coverage of almost anything with people who were to live years later.

Thomas Edison, Figure 1-7. These inventors used their extraordinary abilities to advance the art and science of light-sensitive materials. They envisioned the need for, and value of, recording history as it was happening.

RECREATION

Millions upon millions of photographs are taken each year. Many of them are taken by people who use photography as a hobby. Most people receive much pleasure either by taking pictures or being part of a picture, Figure 1-9.

FIGURE 1-8. Image of an Age—the camera and the automobile grew in popularity together as this 1910 view shows. *Eastman Kodak Company*

FIGURE 1-9. Most people, young and old, enjoy being "captured" in a photograph. *The Morris Co.*

In recent years, manufacturers have engineered cameras that are easy to use. Many cameras are small enough to hold in a person's hand, Figure 1-10. Other cameras, such as the popular 35-mm types, are almost decision-free, Figure 1-11. Most 35-mm cameras have interchangeable lenses and other special features. Microcomputers built into these cameras handle most of the needed calculations in nanoseconds. (A nanosecond is one billionth of a second.) This takes the numerical calculations out of human hands, leaving them up to the preprogrammed computer chips.

Photography can be enjoyed by men, women, and children of all ages. There are no barriers when it comes to squeezing the shutter button. Photo-graphs do not need to be works of art to be meaningful and useful. Photo albums serve as biographies of family and friends. Many people have become biographers by taking pictures over a period of years. The joy associated with recreational photography is unmatched throughout the world.

SUMMARY

Photography touches the lives of all civilized people. The young and old, the weak and strong, and the noneducated and educated rely on photography to bring them needed information. The contents of photographs are as varied as the people who handle the cameras. Photographic film records and remem-

FIGURE 1-10. Small cameras are popular because they can easily be handled and used. *Olympus Corporation*

FIGURE 1-11. A 35-mm camera that only requires an operator to "aim" before taking an accurately exposed picture. *Nikon, Inc.*

bers what the camera lens "sees." This makes photography an extension of the human brain. It is a tool, one of the most useful the world has ever seen.

REVIEW QUESTIONS

Answer these questions to test your knowledge of the unit content.

1. What is the meaning of the word *photography*?
2. T/F Photography has a limited role to play in the overall need for people to communicate.
3. Family _____ are good sources of information about each and every family member.
4. Photography can be considered a universal _____.
5. T/F Words and photographs are equals as recorders of history.
6. Who developed the first practical system of motion pictures?

   A. George Eastman
   B. John Deere
   C. Thomas Alva Edison
   D. Samuel Colt

7. Cameras and _____ grew in popularity together.
8. T/F Photography is today what cave drawings were many years ago.
9. Give two reasons why photography is so popular.
10. Why is photography an extension of the human brain?

AMATEUR AND PROFESSIONAL PHOTOGRAPHY

OBJECTIVES   *Upon completion of this unit, you will be able to:*
- *Define snapshot photography.*
- *Distinguish between amateur and professional photographers.*
- *Explain the difference between freelance and full-time professional photographers.*

KEY TERMS   *The following new terms will be defined in this unit:*
Freelance Photography
Professional Photography
Snapshot Photography

INTRODUCTION

Photography is for everyone. Skill in using cameras and processing equipment is not required. Only limited knowledge about basic cameras is needed before images can be captured on film. Camera and film manufacturers have made every effort to engineer their products for convenience and ease of use.

It is also true that some cameras and photographic equipment require considerable skill and knowledge to operate. Serious amateur and professional photographers have come to expect quality equipment so they can achieve quality in their resulting photographs.

THE SNAPSHOT AMATEUR

Snapshot photographs are those that are taken with little or no planning, Figure 2-1. Also, the cameras used are generally basic and are simple to operate. A film cartridge is inserted into the camera back, the camera is aimed and the shutter is squeezed. No focusing of the lens is necessary owing to the fixed focus lens. This gives the photographer a minimum distance of approximately 6 feet (1.83 m) through infinity in which the subject will be in focus.

Most photographers fall into this general group. People enjoy taking pictures of their families and friends, Figure 2-2. The quality of the result-

ing prints does not match that of professionals, but high quality is not the purpose of taking the picture. The real purpose in taking pictures is to freeze in time those moments that have real meaning at the present.

FIGURE 2-1. Snapshot photographs are taken quickly and without preplanning. *The Holson Company*

FIGURE 2-2. Taking pictures of family and friends is an enjoyable activity.

FIGURE 2-4. An adjustable camera with accessories that permits the serious amateur to take quality photographs. *Pentax Corporation*

FIGURE 2-3. Cameras with a built-in flash make it easy to take pictures in low light. *Olympus Corporation*

Having a personal camera has in recent years become a reality for thousands and thousands of people. This makes it possible to have a camera available at all times. There is no need to schedule the use of the "family" camera as it was once necessary to do. Snapshots can be taken anytime and anywhere. Cameras with a built-in flash make it possible to take pictures in low light, Figure 2-3. The light sensor within the camera determines whether there is need for the flash. This makes it unnecessary for the camera user to even be concerned about light conditions.

THE SERIOUS AMATEUR

People within this group know camera equipment and accessories. They study trade publications and equipment manuals, and purchase sophisticated camera equipment, Figure 2-4. Often, serious amateurs will own two or more cameras and extra equipment that gives them a wide range of possibilities.

Serious amateurs often are found carrying backpacks, Figure 2-5. To the unsuspecting eye, it may appear that the individual is carrying books or possibly lunch. Upon close inspection, it will be found that some very useful and expensive camera equipment and supplies are packed inside, Figure 2-6. This equipment requires attention to detail. Utmost care in handling and using the equipment is the rule and not the exception.

Photographers find interesting locations to take pictures. Some enjoy the out-of-doors in the fields and backwoods, Figure 2-7. Others enjoy the vantage points associated with mountain climbing, Figure 2-8. Wherever it is possible for people to travel and locate themselves, there will be cameras

FIGURE 2-5. Camera bags can be adapted with a harness to be carried on the back. *Tenba, Inc.*

FIGURE 2-6. The serious amateur photographer fills the backpack with cameras, extra lenses, flash unit, film, and other useful accessories. *Tenba, Inc.*

FIGURE 2-7. Interesting photographs of plants, bushes, and trees are pursued and recorded on film. *Sports & Photo*

there at the same time or shortly thereafter. Serious amateur photographers have produced, and continue to produce, excellent photographs.

THE FREELANCE PROFESSIONAL

People in this group receive monetary reward for their efforts. Often, this type of photographer will accept an assignment from a company to take a series of photographs about its employees, Figure 2-9. Once taken, the film is processed, and prints are made. Frequently, the freelance photographer will have a personal darkroom that includes an

FIGURE 2-8. Location is seldom a problem for the serious amateur. *Tenba, Inc.*

FIGURE 2-9. Employee photographs often are taken by freelance professional photographers. *Santa Fe Railway*

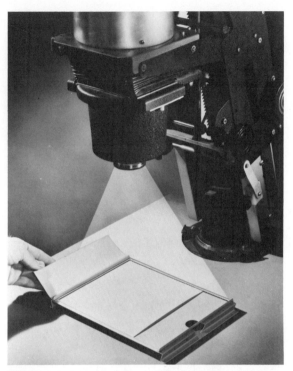

FIGURE 2-10. An enlarger permits the freelance photographer to produce prints quickly and efficiently.

enlarger, Figure 2-10. This permits quick turnaround time and personal service.

Maintaining full employment outside professional photography is typical of freelance professionals. Also, a number of people are full-time, freelance photographers. This means that they are totally self-employed. Through self-motivation, they seek short-term contracts with companies and individuals. For example, they might prepare a series of photographs that can be used in printed promotional brochures or a slide series that will be shown to interested groups of people.

Many freelance professionals are masters at organizing a series of photographs or slides that will later be sold to magazines or advertising agencies. Often, a photograph or slide is easiest to sell when the results can be viewed and studied by the purchaser. Flowers make good subjects for a colorful slide series, Figure 2-11. A special lens and an electronic flash make it possible for close-up exposures to be taken.

FIGURE 2-11. Taking pictures of flowers often provides a challenge that the freelance photographer enjoys. *Vivitar Corporation*

FIGURE 2-13. Written scripts are often part of a professional slide series. *Eastman Kodak Company*

FIGURE 2-12. A lighted slide sorter is a useful device for selecting the best images. *Sima Products Corporation*

After a series of slides are taken, they must be organized. Also, the best exposures need to be selected. A useful piece of equipment for this task is a lighted slide sorter, Figure 2-12. The unit contains channels for the slides to rest in and a magnifying glass for viewing selected portions of an image. At this point, the photographer may not be finished. The next step could be to write a script that will accompany the slide series, Figure 2-13.

This can take considerable time and a concentrated amount of library research.

THE FULL-TIME PROFESSIONAL

Earning a living is the primary objective of a full-time professional photographer. It is necessary that attention be paid to every detail. Professional photographs must give every appearance of quality, otherwise the professional will not stay in business.

A typical example of a professional portrait is shown, Figure 2-14. The subject was positioned to display his best facial features. The lighting provides highlights that draw viewer attention. Another professional example shows the comfortable seating in a passenger plane, Figure 2-15. The photographer instructed each person to do something specific that emphasized their comfort. Establishing and controlling the scene is an important part of professional photography.

SUMMARY

Photography has come a long way since its beginnings in the middle of the 19th century. In the past, it was cumbersome to take a picture and then to

FIGURE 2-14.   This professional portrait is designed to emphasize the strengths of a corporate vice president. *Berkey Marketing Companies*

FIGURE 2-15.   A photograph taken inside a commercial airplane is used to show passenger comfort. *American Airlines*

process the results. Images were poor according to contemporary standards. Today, people with all levels of ability can enjoy taking pictures. Cameras requiring very little skill are available to nearly everyone. Skill and creativity are necessary attributes for the professional photographer. Where there are people, there will be photography.

REVIEW QUESTIONS

Answer these questions to test your knowledge of the unit content.

1. Why is photography for everyone?

2. Pictures taken with little to no preplanning are called _____.
3. T/F Cameras used to take the majority of snapshots usually have fixed focus lenses.
4. T/F Personal cameras were commonplace for people during the first half of the 20th century.
5. Serious amateur photographers are often seen carrying their cameras, accessories, and supplies in _____.
6. T/F Geographic location discourages photographers from taking pictures with their expensive equipment.
7. The main characteristic that distinguishes the amateur from the professional is:

   A. Quality of equipment
   B. Money for services
   C. Praise and recognition
   D. General appearance

8. T/F Freelance photographers may or may not have employment outside their photography business.
9. When is a photograph easiest to sell?
10. What is the primary objective of a professional photographer?

PHOTOGRAPHIC SCIENCE AND TECHNOLOGY

OBJECTIVES   *Upon completion of this unit, you will be able to:*
- *Recognize how and why photography is used in medical science.*
- *Describe the technology of high-speed photography.*
- *Note the use of photography in the study of celestial bodies.*

KEY TERMS   *The following new terms will be defined in this unit:*

High-speed Photography	*Photomicroscope*
Photomicrograph	*X-ray Photograph*

INTRODUCTION

Science and technology rely heavily on photography. Film is used to "save" images created and observed by scientists. Later, intense study of these images can be conducted. Cameras can be attached to telescopes, microscopes, and other sophisticated equipment. In a sense, the camera, with its ultra-sensitive film, becomes an extension of the human mind.

MEDICAL PHOTOGRAPHY

Many hospitals have medical photographers on their in-house staffs. Hospitals that do not provide this service have standing contracts with private photographers who have agreed to take pictures at a moment's notice. Without photography, physicians would be unable to use enough words to describe the appearance of a specific disease or a wound. This was discovered early in the use of photography, Figure 3-1.

There are two primary purposes for medical photography: (1) to document unusual medical procedures or rare cases of diseases or injuries, and (2) to maintain visual evidence in case of law suits. The latter is especially true with plastic surgery. An example of how photography is used for documentation purposes is in pathology, Figure 3-2. Pathology is the science of interpreting and diagnosing the changes caused by disease in living tissue. Portions of the animal or human body are photographed for both study and evidence.

Photography is useful as a diagnostic tool in medical science. Cameras are fitted to microscopes, allowing researchers to record on film what can be seen through the eyepieces. One example is shown, Figure 3-3. The camera is attached to the top of the "photomicroscope." This provides a direct line of sight to the specimen plate at the base, Figure 3-4. A *photomicroscope* is an instrument that combines a microscope, camera, and light source. It is used to

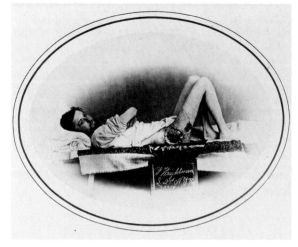

FIGURE 3-1. Photography found a place in medical technology during the United States Civil War (1861–1865). This photograph clearly shows a gunshot wound of the hip. *Armed Forces Institute of Pathology*

FIGURE 3-2. Photography is useful for retaining medical images in the pathology laboratory. *Department of Pathology, Louisiana State University Medical Center*

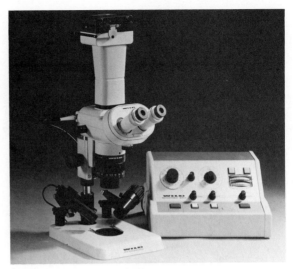

FIGURE 3-3. A photomicroscope with a camera affixed to the top of the instrument. *E. Leitz, Inc.*

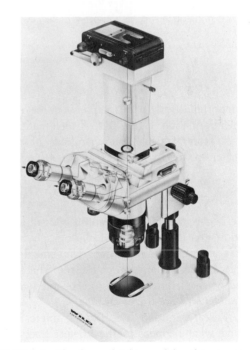

FIGURE 3-4. The lines-of-sight used by the viewer and camera in the photomicroscope shown in Figure 3-3. *E. Leitz, Inc.*

produce *photomicrographs*. These are photographs of a magnified image of a small object.

Cameras are built inside some photomicroscopes, Figure 3-5. The 35-mm camera in this model is completely integrated into the instrument, Figure 3-6. This system design allows perfect control of the images prior to exposing the film.

Most people are familiar with the frequent use of *x-ray photographs* by physicians. These filmed images permit diagnostic work prior to surgery or medical treatment. Cameras have been designed to take photographs inside the body. This too has reduced the need for surgery.

In many ways, photography is helping to meet the needs of people who require medical services. Medical science relies heavily on photography in more ways than can be shown in this book.

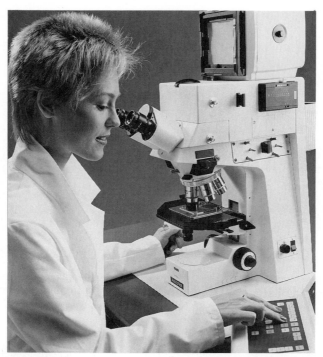

FIGURE 3-5. A photomicroscope containing a fully integrated camera and computer control system. *Carl Zeiss, Inc.*

One important use of photography is in the area of patient information, Figure 3-7. A well-made slide series is more valuable than hundreds of words written or spoken by medical personnel.

HIGH-SPEED PHOTOGRAPHY

People began attempting to capture movement on film soon after the invention of photography. Most cameras are capable of stopping movements that take place quicker than the eye can see. This is evidenced when pictures are taken of people walking. The movements are frozen in time.

Researchers and scientists are interested in knowing what happens in the briefest fraction of a second. Standard cameras could not meet the requirements; thus, special high-speed cameras were designed, Figure 3-8. High-speed cameras use long rolls of film and take individual pictures at very high rates of speed. A typical use for high-speed cameras is shown, Figure 3-9. The photograph made it possible to study the accident long after it took place.

CELESTIAL PHOTOGRAPHY

Close up views of the moon, stars, sun, and other planets have long been a fascination of people throughout the earth. Telescopes made it possible to see many of the celestial sights. Soon cameras were fastened to telescopes, making it possible to take pictures of far off planets, Figure 3-10. Scientists could now critically evaluate the aspects of a planet long after the photograph was made.

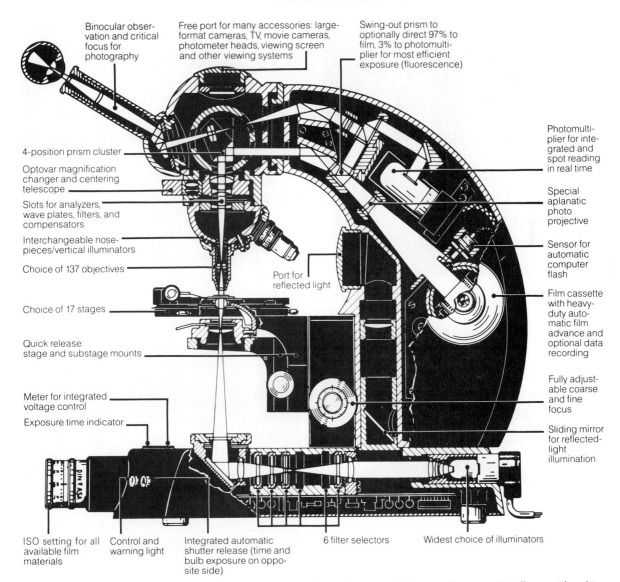

Binocular obser-
vation and critical
focus for
photography

Free port for many accessories: large-
format cameras, TV, movie cameras,
photometer heads, viewing screen
and other viewing systems

Swing-out prism to
optionally direct 97% to
film, 3% to photomulti-
plier for most efficient
exposure (fluorescence)

4-position prism cluster

Optovar magnification
changer and centering
telescope

Slots for analyzers,
wave plates, filters, and
compensators

Interchangeable nose-
pieces/vertical illuminators

Choice of 137 objectives

Choice of 17 stages

Quick release:
stage and substage mounts

Meter for integrated
voltage control

Exposure time indicator

Port for
reflected light

Photomulti-
plier for inte-
grated and
spot reading
in real time

Special
aplanatic
photo
projective

Sensor for
automatic
computer
flash

Film cassette
with heavy-
duty auto-
matic film
advance and
optional data
recording

Fully adjust-
able coarse
and fine
focus

Sliding mirror
for reflected-
light
illumination

ISO setting for all
available film
materials

Control and
warning light

Integrated automatic
shutter release (time and
bulb exposure on oppo-
site side)

6 filter selectors

Widest choice of illuminators

FIGURE 3-6. A schematic view of a photomicroscope. The application of photography was critically considered in the design of this precision instrument. *Carl Zeiss, Inc.*

Photographs of the moon have become very popular ever since man first set foot on the earth's only known natural satellite, Figure 3-11. Prior to the first walk on the moon by Neil A. Armstrong on July 20, 1969, many photographs had been taken. These gave scientists and astronauts information that was vitally needed for their successful mission. Cameras have been taken on every space venture,

FIGURE 3-7.   Photography is a useful tool for showing the medical services of a hospital. *Eastman Kodak Company*

FIGURE 3-9.   A high-speed camera makes it possible to take fast-action photographs like this example. *Instrumentation Marketing Corporation*

FIGURE 3-8.   A high-speed camera capable of taking 24 to 500 frames per second. *Instrumentation Marketing Corporation*

FIGURE 3-10.   A picture of the planet Saturn taken with the benefit of a telescope. *Celestron International*

SUMMARY

Photography has and will continue to be heavily used to advance science and technology. Other types of photography are holography, infrared, and ultraviolet. Photography is used for surveillance, security, law enforcement, military and industrial

providing thousands of photographs for experts to study as man continues to explore and invade space.

FIGURE 3-11. Photographs of the moon provided much information prior to the landing of men on its surface. *Celestron International*

FIGURE 3-12. A night vision camera system. This permits night vision viewing and picture taking.

documentation. A camera attached to a special "night vision" scope permits pictures to be taken in the dark, Figure 3-12. Other uses for scientific and technological photography are left to the imaginations of people who wish to become involved.

REVIEW QUESTIONS

Answer these questions to test your knowledge of the unit content.

1. Images are _____ on film for later scientific and technological study.
2. Explain what is meant by the statement, "The camera is an extension of the human mind."
3. T/F All hospitals have full-time photographers on their staffs.
4. Name the two primary purposes of medical photography.
5. A photomicroscope can have a camera:
   A. Attached to the top
   B. Built inside
   C. Attached either to the left or right side
   D. All of the above answers
6. T/F A good method to present general medical care to patients is through the use of slides.
7. T/F High-speed cameras use individual sheets of film.
8. What type of people are interested in high-speed photography?
9. T/F Celestial photography is a new science that has developed only in the last 20 years.
10. T/F Photographs can be taken in the dark of the night with special equipment and film.

THE PHOTOGRAPHIC INDUSTRY

OBJECTIVES   *Upon completion of this unit, you will be able to:*
- *Discuss the scope and size of the photography industry.*
- *Identify the three major amateur market areas.*
- *Explain the general operation of commercial film and print processing laboratories.*
- *Recognise the importance of photographic supplies and equipment manufacture.*

KEY TERMS   *The following new terms will be defined in this unit:*
Commercial Processing Laboratory   Quick Photographic Processing
GNP   Photofinishing

INTRODUCTION

Photography is big business. It touches the lives of virtually all human beings. Because of this, a vast number of manufacturing companies have been created since the beginning of amateur and commercial photography in the late 1800s. Sales outlets and service centers also have become very strong.

SCOPE AND SIZE

Photographic film and general supplies are available to the amateur photographer almost anywhere, Figure 4-1. Black-and-white and color film in different sizes can be obtained in department stores, grocery stores, pharmacies, and service stations. Also, film products are available through vending machines, catalog outlets, discount stores, and full-time camera and film businesses.

During the first half of the 20th century, people became interested in taking pictures of people, places, and things. Unfortunate economic and political conditions throughout the world did not permit a wide-spread expansion of the industry. Most people were too involved in making a living and protecting their freedoms to spend much time with photography. Those people who did have the time and money made good use of photography. Cameras were manufactured to meet the needs of

the public just prior to the Great Depression of the 1930s, Figure 4-2.

The second half of the 20th century brought much prosperity to people, and companies engaged in one or more phases of the photographic industry. Economists consider photography to be a dramatic growth industry. Technological advancements have fueled the growth, Figure 4-3. Taking pictures has been made easy. With the increase in disposable personal income, people are willing to spend money on photography. The actual growth of the photographic GNP continues to exceed the growth of the United States *GNP* (Gross National Product).

FIGURE 4-1. Film and general supplies for the amateur photographer can be obtained in many locations.

FIGURE 4-2.   Cameras being assembled by skilled workers at the Kodak Camera Works in 1929. *Eastman Kodak Company*

FIGURE 4-3.   Cameras designed for picture-taking convenience have created a large market for the photographic industry. *Eastman Kodak Company*

AMATEUR MARKET AREAS

The wide-spread amateur photography market can be divided into three major segments: (1) cameras and auxiliary equipment, (2) film and general supplies, and (3) photofinishing. The growth of these three segments has been substantial in recent years, Figure 4-4. Combined sales in a recent year exceeded $7 billion dollars. Another important marketing fact is that photography continues to demonstrate its recession-proof characteristics.

Cameras and auxiliary equipment are purchased by people who are interested in upgrading their picture taking capabilities, Figure 4-5. Because of this interest, they shop at different stores and sales outlets until they find the camera, flash unit, or extra lens that will fit their needs. This segment of the amateur market alone accounts for a vast number of businesses engaged in retail photography sales.

The photofinishing segment includes film and print processing. It is a large part of the amateur photographic market. *Processing laboratories*

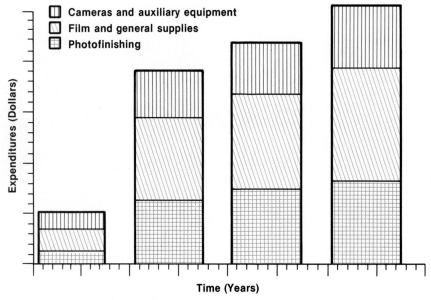

▯ **Cameras and auxiliary equipment**
▯ **Film and general supplies**
▯ **Photofinishing**

Expenditures (Dollars)

Time (Years)

FIGURE 4-4. Photography has enjoyed a dramatic growth in recent years.

FIGURE 4-5. People are frequently interested in upgrading their photographic equipment.

have established a variety of retail locations where amateur photographers can leave their exposed film, Figure 4-6. The retail store sends the film to the laboratory for processing via the mail or a courier service. The common procedure is for the retail store to ship the accumulated film on designated days of the week. In turn, they will receive delivery of processed film and prints or slides on the same days. Generally, the turn-around time is 7 days or less.

COMMERCIAL PROCESSING

Professional photographers establish open accounts with processing laboratories. They often mail their exposed film directly to the laboratory. Also, they may use a courier service provided by the commercial laboratory. Private citizens also may mail their film directly to the laboratory by using a special mailing envelope, Figure 4-7. Some mailing envelopes must be purchased. The purchase price covers the cost of film developing, print making, or slide mounting. With other mailing envelopes,

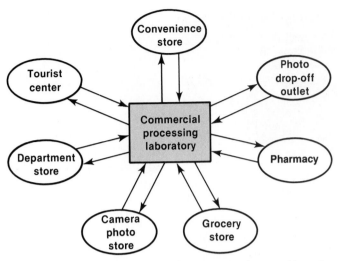

FIGURE 4-6. Commercial laboratories receive film for processing from a wide variety of retal stores.

FIGURE 4-7. Mailing envelopes make it convenient for amateur photographers to send film to the laboratory for processing. *Mystic Color Lab*

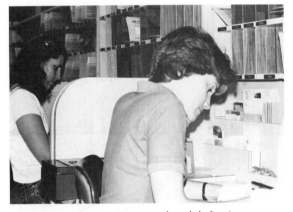

FIGURE 4-8. Organization, speed, and dedication are necessary when handling several thousand film and print orders on a daily basis. *Mystic Color Lab*

payment must be included with the exposed roll or disk of film.

Commercial laboratories must be efficient in handling film, prints, and slides from their custom-

ers. Organization is the key to making 250,000 prints in a single working day, Figure 4-8. Also, production equipment that is computer controlled is essential to consistent quality, Figure 4-9. People are still very important in film and print processing laboratories. They not only handle paperwork, set up machines, and process orders, but qualified people serve as inspectors, Figure 4-10. The human eye is the final judge as to whether a print has been made correctly.

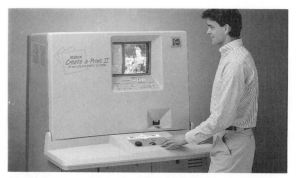

FIGURE 4-11. Create-a-Print enlargement equipment can be easily operated by amateur photographers who wish to produce "quick" color prints from color negatives. *Eastman Kodak Company*

FIGURE 4-9. Computer-controlled equipment makes it possible to match processed film and prints. *Mystic Color Lab*

FIGURE 4-12. Much of the minilab/quick processing equipment is self-contained and fully automated. *Noritsu America Corporation*

FIGURE 4-10. Skilled personnel visually inspect every print in a commercial processing laboratory. *Mystic Color Lab*

QUICK PHOTOGRAPHIC PROCESSING

Photographic print creating equipment has made it possible for amateur photographers to create color prints in less than five minutes, Figure 4-11. Convenient locations and the fast turn-around time are features that appeal to amateur photographers. Equipment that makes quick processing possible is highly automated, Figure 4-12. It is also possible to locate the equipment in a lighted room. The only time care must be exercised is when film is loaded into the film processing unit. Once film is inside the machine, the entire processing from film to prints is done within the system.

MANUFACTURING SUPPLIES AND EQUIPMENT

A large part of the photographic industry involves the manufacture of supplies and equipment. Every piece of equipment—cameras, flash units, lenses, tripods, processors, etc—must be engineered and marketed. This is also true with film, photographic paper, and the many other supplies associated with all types of photography. Assembly of precision equipment must be done by highly skilled people, Figure 4-13. Photographic equipment includes some of the most sophisticated equipment available. Inspecting a camera quickly reveals the engineering and manufacturing standards that are necessary to produce quality equipment.

SUMMARY

Photography has captured the interest of millions of people. Pictures are being taken every second of every day. This has made the photography industry a giant among all industry. Amateur and professional photographers use great amounts of film, paper, and supplies. They also use large quantities of large and small equipment. The photography industry continues to meet the demands of both the individual and large corporations.

REVIEW QUESTIONS

Answer these questions to test your knowledge of the unit content.

1. Why has photography become part of big business?
2. Name five types of retail stores that handle photographic film.
3. T/F Camera manufacture has been fully automated since the beginning of the 20th century.
4. Photography is a _____ industry.
5. List the three segments of the amateur photographic market.

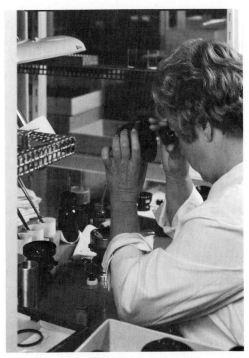

FIGURE 4-13. The assembly of lenses must be done with precision and care. *Optische Werke G. Rodenstock*

6. T/F Retail stores deliver film to processing laboratories on a scheduled basis.
7. Private citizens can deliver a roll of exposed film directly to a processing laboratory by using a special _____ _____.
8. The _____ _____ is the final judge of print quality in a commercial processing laboratory.
9. Quick photographic processing facilities are frequently located in:

   A. Shopping centers   C. Small towns
   B. Grocery stores   D. Universities

10. T/F The manufacture of supplies and equipment is a major segment of the total photographic industry.

PHOTOGRAPHIC CAREER OPPORTUNITIES

OBJECTIVES   *Upon completion of this unit, you will be able to:*
- *Summarize aspects of being a photographer.*
- *Describe the work of a photofinisher.*
- *Discuss photography requirements in audiovisual programming, research, and quality control.*
- *Prepare a portfolio of photography samples.*

KEY TERMS   *The following new terms will be defined in this unit:*

Photojournalist   *Photoretoucher*

Photo minilab   *Portfolio*

INTRODUCTION

Cameras, enlargers, film processors, and other photographic equipment are only tools. Preparation for a career in photography requires strong development in the many technical proficiencies, but more important is the development of the mind to see photographs in advance of taking the picture. Also, the photographer must understand what is happening or what could happen. This makes it possible to record the scene as accurately and completely as it really is. Successful photographers must improve their minds through study, research, reading, and personal observation.

BEING A PHOTOGRAPHER

In 1900, an advertisement appeared in newspapers and magazines encouraging people of all ages to become photographers, Figure 5-1. This campaign continued the advertising program started in the late 1800s to promote photography for everyone. Today, photography is very popular with almost everyone, as evidenced by the millions of pictures taken by amateur photographers.

Photography is the tenth largest industry in the United States according to recent information released by the government. This indicates strong employment opportunities for competent photog-

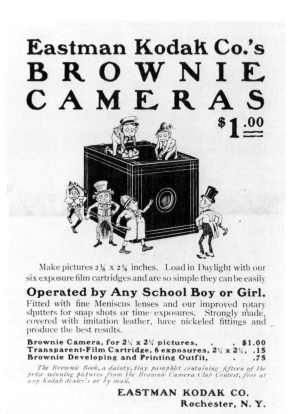

FIGURE 5-1. The Brownie camera placed photography within the financial reach of almost everyone. This camera may have encouraged many young people to make photography a career. *Eastman Kodak Company*

raphers. Currently, there are about 100,000 photographers employed throughout the country. Many photographers are involved in portrait photography, Figure 5-2. People of all ages are interested in having a formal portrait taken of themselves. Portrait photographers are often in business for themselves and are located in large and small cities. Frequently, these are husband-and-wife team businesses.

Photojournalists deal with newsworthy events, people, and places, Figure 5-3. These people work for publications such as newspapers, magazines, and television news shows. Law enforcement also has a strong need for qualified photographers. People who work in this specialty area need to be familiar with many kinds and sizes of cameras, Figure 5-4.

FIGURE 5-3. Photojournalists often use special equipment to photograph events that will be reported in various publications. *Celestron International*

FIGURE 5-2. Portrait photography is a career choice for many talented people. *Photo Control Corporation*

FIGURE 5-4. Photographers who work in law enforcement must be familiar with the smallest of cameras. *Karl Heitz, Inc.*

Creativity and excitement provide strong incentives for this growing area for photography.

Both still and motion picture photographers are needed who have proper training and good backgrounds. These photographers frequently are sent on assignment to interesting locations, Figure 5-5. This gives them the opportunity to work with and around people of other nationalities. Photographers need to possess skill in using cameras. They also must be able to compose pictures with creativity and recognize a potentially good photograph.

PHOTOGRAPHIC FINISHING

Professional photographers frequently rely on workers in photofinishing laboratories to develop their film and make photographic prints, Figure 5-6. Darkroom technicians must be able to do everything necessary to develop and print film. Special skills are needed to meet some of the professional photographer requirements. This is especially true for the needs involving retouching photographs and negatives, Figure 5-7. *Photoretouchers* must be artists who know the proper use of tools and materials used in retouching work. They also must have excellent eyesight, steady nerves, and a sense for knowing how to achieve quality.

Amateur photographers also make much use of film processing laboratories. With the new films and changes in film formats, special requirements must be met by the photofinisher, Figure 5-8. Special equipment must be purchased to handle the rolls and disks of film.

One-hour film and print processing laboratories have opened many career opportunities in

FIGURE 5-6.   A laboratory technician adding chemistry to a film processor. *Colenta America Corporation*

FIGURE 5-7.   Photoretouchers must possess special artistic skills. *Mystic Color Lab*

FIGURE 5-5.   Photographers often have interesting assignments. *Instrumentation Marketing Corporation*

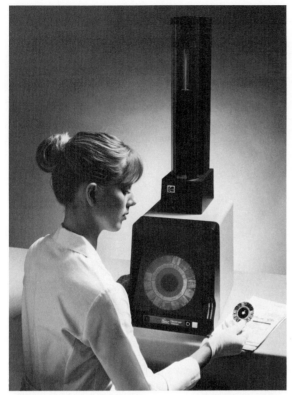

FIGURE 5-8. Photofinishers must carefully handle film to keep it clean and free of scratches. *Eastman Kodak Company*

FIGURE 5-9. Photo minilabs provide many employment opportunities.

FIGURE 5-10. Audiovisual photographers often take slides according to a prepared script.

recent years, Figure 5-9. These *photo minilabs* are located in population centers, such as shopping centers and busy parts of a city. Managers and technical personnel will be needed for these installations for years to come. In the photofinishing area, there are about 80,000 employed personnel. Employment opportunities will continue to increase in this area.

AUDIOVISUAL PROGRAMMING

People in this area of photography are required to work in a variety of areas. Audiovisual (A-V) photographers take pictures of a wide range of subjects. They may do photography for company promotionals, advertising agencies, educational programs, or government groups, Figure 5-10. When the

FIGURE 5-11. Photographic technicians are needed to program and operate projection equipment. *Eastman Kodak Company*

FIGURE 5-12. A photograph taken in 1878 by George Eastman. *Eastman Kodak Company*

slides have been taken, there is a need to organize the entire series of slides according to the prepared script.

Programming requirements are fulfilled by people who are skilled in operating and controlling equipment, Figure 5-11. Computer skills are necessary to synchronize slide projectors for a multi-image presentation. Companies and corporations keep the demand high for people with these skills.

RESEARCH AND QUALITY CONTROL

A 24-year-old banker pointed his camera at the house across the street, Figure 5-12. This took place on a snowy day in 1878 in Rochester, New York. The young man was George Eastman, and his hobby was photography. The quality of his research results encouraged him to turn his hobby into a business career. The rest is history for today the Eastman Kodak Company is the largest producer of photographic equipment and products in the world.

Scientific photographers provide illustrations and documentation for scientific publications and research reports, Figure 5-13. Medical researchers often use ultraviolet and infrared photography to obtain needed information. Pathologists often take

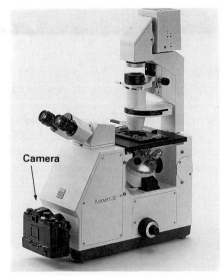

FIGURE 5-13. Sophisticated equipment, such as this photomicroscope, is necessary for scientific photographers. *Carl Zeiss, Inc.*

pictures of body organs so they can research and study a particular disease, Figure 5-14.

Quality control experts are needed in every phase of photography equipment and supply manufacture, Figure 5-15. These people must understand what each component does. They must be able to analyze problems when they occur. Also, quality control personnel must anticipate difficulties before they occur. Increases in automation have created the need for highly educated and dedicated workers in this area of photography.

YOUR PORTFOLIO

A *portfolio* is a collection of work placed in a case with a hinged cover. The case should be large enough to hold prints up to 11 × 14 to 16 × 20 inches (28 × 35.6 to 40.6 × 50.8 cm) in size.

A wide sampling of work should be included in the portfolio collection. Black-and-white prints, color prints, and color slides are needed for a complete representation of the types of final photographic products. Within these three groups, it is important to include several photographic examples.

EDUCATION AND EXPERIENCE REQUIREMENTS

Photography has no set entry requirements for formal education or training. Employers usually seek applicants who have a broad technical understanding of photography, as well as other photographic talents, such as imagination, creativity, and a good sense of timing. Technical expertise can be obtained through practical experience, postsecondary training, or some combination of the two. Some jobs require that applicants have specialized knowledge in areas outside of photography.

Photographic training is available in colleges, universities, junior colleges, and art schools. Several colleges and universities offer 4-year curriculums leading to a bachelor's degree in photography.

FIGURE 5-14. Pathology laboratory technicians use photography to record pictures of diseased body organs. *From SPECIMEN DISSECTION AND PHOTOGRAPHY FOR THE PATHOLOGIST, ANATOMIST AND BIOLOGIST, by M. Donald Mcgavin and Samuel Wesley Thompson.* Courtesy of Charles C. Thomas, Publishers, Springfield, Illinois.

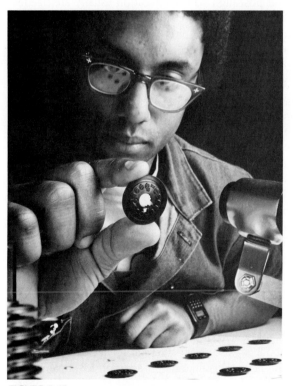

FIGURE 5-15. Quality assurance personnel are important in the manufacture of standardized equipment and supplies. *Eastman Kodak Company*

Some colleges and universities grant master's degrees in specialized areas, such as photojournalism. In addition, some colleges have 2-year curriculums leading to a certificate or an associate degree in photography. The Armed Forces also train people in photographic skills. A formal education in photography gives a fundamental background in a variety of equipment, processes, and techniques.

A person may prepare for work as a photographer in a commercial studio through 2 or 3 years of on-the-job training as a photographer's assistant. Trainees generally start in the darkroom where they learn to mix chemicals, process film, and do photoprinting. Later, they may set up lights and cameras or help an experienced photographer take pictures.

Amateur experience is helpful in getting an entry job with a commercial studio, but high school and post-high school education usually are needed for industrial or scientific photography. In these areas, success in photography depends on being more than just a competent photographer. Adequate career preparation requires some knowledge of the field in which the photography is used. For example, work in scientific, medical, and engineering research, such as photographing microscopic organisms, requires a background in the science or engineering specialty, as well as skill in photography.

Photographers must have good eyesight and color vision, artistic ability, and manual dexterity. They should be patient, accurate, and enjoy working with detail. Some knowledge of mathematics, physics, and chemistry is helpful for understanding the use of various lenses, films, light sources, and development processes.

EMPLOYMENT OUTLOOK

Employment of photographers is expected to grow about as fast as the average for all occupations in the next several years. In addition to openings resulting from increased demand for photographers, others will occur each year as workers transfer to other occupations, retire, or die.

Employment is expected to grow as business and industry place greater importance upon visual aids in meetings, stockholders' reports, sales campaigns, and public relations work. Photography is becoming an increasingly important part of law enforcement work, as well as scientific and medical research, where opportunities are expected to be good for those with appropriate technical skills. Employment in photojournalism is expected to grow slowly.

Portrait and commercial photography is expected to grow slowly, and competition for jobs as portrait photographers and photographers' assistants is expected to be keen. These fields are relatively crowded since photographers can go into business for themselves with a modest financial investment or work part time while holding another job.

SUMMARY

Employment possibilities for photographers and photofinishers will continue to increase in the years ahead. Photography offers opportunities for people with a wide range of interests and abilities. There will always be a need for people to take pictures. The methods, equipment, and materials may change, but people will continue to ask for visual images of themselves, friends, objects, and their total environment. Anyone wishing to pursue photography as a career must remember and do the following: be professional, be reliable, be sensitive, be creative, be competent, and be a quality person in every way possible.

REVIEW QUESTIONS

Answer these questions to test your knowledge of the unit content.

1. Is technical proficiency the only requirement for being a photographer? Explain.
2. T/F The high cost of photography even in the early years has kept people from becoming involved.

3. How large is the photography industry in the United States?

    A. 5th   C. 12th
    B. 10th   D. 20th

4. About how many photographers are currently employed in the United States?

    A. 10,000   C. 100,000
    B. 50,000   D. 250,000

5. T/F Law enforcement uses photography to great advantage.

6. Which type of processing facility is often found in shopping centers?

7. How are computer skills used in audiovisual photography?

8. Who turned a hobby into a giant business?

9. T/F Serious photographers need to prepare and maintain portfolios of their work.

10. List the six requirements for a career as a photographer.

CHAPTER TWO

CAMERAS FOR PHOTOGRAPHY

CAMERA BASICS

OBJECTIVES   *Upon completion of this unit, you will be able to:*
- *Describe the relationship of a camera to the human eye.*
- *Identify the basic parts of a camera.*
- *Name the four categories (groups) of cameras.*
- *Care for and handle cameras in a proper manner.*
- *Discuss historical highlights of cameras.*

KEY TERMS   *The following new terms will be defined in this unit:*

Adjustable Camera	Iris	Single-lens Reflex Camera
Aperture	Iris Diaphragm	Twin-lens Reflex Camera
Camera Obscura	Lens	View Camera
Disc Camera	Pupil	Viewfinder
Film Holder	Retina	Viewfinder Camera
f-stop Numbers	Shutter	

INTRODUCTION

Thousands of cameras of many shapes and sizes are in use today. Millions of people in all parts of the world own and use cameras often. Most people do not think much about a camera's history, manufacture, or the many parts needed to make it work. This unit and other units in this book will treat these plus many other important points about simple and complex cameras.

THE CAMERA

The human eye is a sophisticated living camera. Other living cameras are essential to the lives of animals, fish, birds, and even insects. These living cameras capture images through the nerves of the eye and send the information to the brain where the visible image becomes meaningful.

A camera is a mechanical-electronic device designed and manufactured by humans. It is a tool or a piece of equipment designed to accurately control the amount of light that reaches the light-sensitive film. A camera and an eye are much the same in that light may enter only through a con-trolled opening, Figure 6-1. In the eye, this opening is called the *iris*, and in a camera, it is called the *aperture*. This controlled opening on cameras is accurately measured and adjusted using *f-stop numbers* (see unit 14). The *pupil* of the eye and the lens of the camera gather the reflected light and focus it into a sharp image. The *retina* and the film, both sensitive to light, record the controlled image with precision.

BASIC CAMERA PARTS

All cameras contain the same basic parts. The major difference from one camera to another is the location of certain parts. This is especially true with *adjustable cameras*, such as those presented in units 9, 10, and 11.

Becoming familiar with these basic camera parts is important, Figure 6-2. It then will be possible to learn and use the additional parts and controls on more advanced cameras.

BODY.   The body is a very important part of any camera. It must be rigid and of sound construction. It has to be designed for convenience in handling, but it must be large enough to hold the

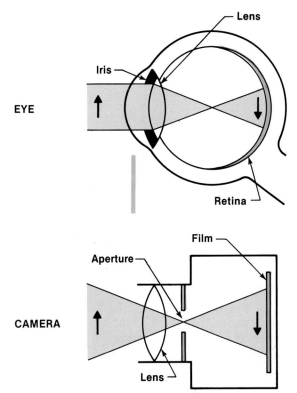

FIGURE 6-1. The eye and the camera obtain pictures in similar ways.

various camera supports and controls. Camera bodies are made from plastic, wood, and metal.

LENS. This is the "eye" of a camera. Its function is to gather light and focus it on the film. The light passing through a camera lens forms the visible image on the film after it is processed. The illustration shows only one lens element, but nearly all camera lenses contain several single elements forming a "lens system" (see unit 13). Camera lenses usually are made from high quality glass, but plastic lenses are found in economical cameras.

VIEWFINDER. The *viewfinder* allows the camera user to view the scene that will be captured on film. Several types of viewfinders are found on cameras of different styles and quality. Some are very simple and others are complex, but all are designed to permit the camera user to aim the camera toward the selected scene.

FILM HOLDER. Cameras are designed to use film that is packaged in single sheets, rolls, cartridges, and cassettes. This variety of film formats requires many different styles of film holders. A specific camera is designed to use one film format. The film holder must be made to control the film location in a precise manner in cameras of all styles and sizes.

FILM ADVANCE. This is a mechanism that moves the film forward so a new picture can be taken. The film advance must be very precise so that the right amount of film is advanced each time. Cameras contain film advances that are operated by hand, spring, or electric motor.

APERTURE. The *aperture* is the opening that is used to control the amount of light that will strike the film. Simple cameras contain a fixed-sized aperture opening while adjustable cameras use adjustable apertures. These are made of thin metal leaves that overlap. An adjustable aperture is often called an *iris diaphragm*. Specific aperture openings are identified by f-stop numbers (see unit 14).

SHUTTER RELEASE. This device is usually positioned on the top right side of most cameras. Pushing or squeezing it activates (open and closes) the shutter so a picture can be taken. The shutter is similar to the eyelid of the human eye: it is either open or closed. As with the aperture, simple cameras contain a fixed-speed shutter, while adjustable cameras use shutters that can be regulated for different speeds.

CAMERA CATEGORIES

There are four photographic camera groups or categories. Within each group, there are specific subcategories of camera types, brands, and models. Over the years, manufacturers throughout the world have designed and built hundreds of different cameras. Each and every camera design has one main purpose—to capture images either on photographic film or with electronic material (see unit 44).

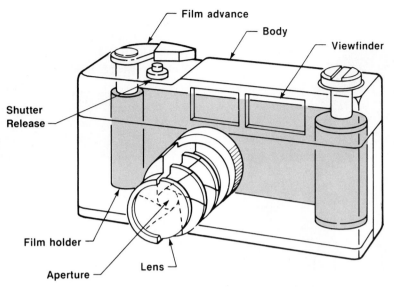

Film advance

Body

Viewfinder

Shutter
Release

Film holder

Aperture

Lens

FIGURE 6-2. All cameras must have these seven basic parts.

Some cameras are small and others are large; some are simple, while some are complex. Never forget that it takes human skill and attention to detail for quality photographs. Expensive cameras, accessories, and film do not guarantee quality. It is possible to obtain beautiful photographs using cameras of low to medium price.

Consistent photographic quality often is achieved only when cameras are of good design and construction. This is also true of other precision equipment, such as a wristwatch or even a computer. Purchasing brand name cameras and other photographic equipment is generally very wise. The cost may be more, but the value and long-range benefits can be much greater.

VIEWFINDER CAMERA. The most basic camera category is that known as the *viewfinder.* These cameras are available in many models and sizes. A popular camera style that can be used by nearly anyone is shown in Figure 6-3. Another popular camera for the amateur is one that uses roll film cartridges, Figure 6-4. Most viewfinder cameras use fixed focus, fixed aperture, and fixed shutter speed.

FIGURE 6-3. An entry-level 35mm camera that can easily be operated by the beginning photographer. *Eastman Kodak Company*

This permits ease of use for all photographers—from the very young to the very old and from the novice to the professional (see unit 8).

SINGLE-LENS REFLEX CAMERA. The camera group that has become very popular in recent years is the *single-lens reflex,* Figure 6-5. These are often

FIGURE 6-4.   Amateur photographers can easily use cameras of this style. *Eastman Kodak Company*

FIGURE 6-5.   The single-lens reflex camera is very popular with serious photographers. *Nikon, Inc.*

FIGURE 6-6.   Manufacturers make it possible to obtain various lenses and viewfinder systems on many single-lens reflex cameras. *Pentax Corporation*

called SLR cameras. Many different companies manufacture and sell these types of cameras in all parts of the world. Adjustments such as aperture opening, shutter speed, and film speed are common to this camera group. Many single-lens reflex cameras have a wide variety of lens and viewing systems available, Figure 6-6. This feature makes the camera useful for nearly all kinds of photography. (see unit 9).

  TWIN-LENS REFLEX CAMERA. *Twin-lens reflex cameras* have the combined features of viewfinder and single-lens reflex cameras, Figure 6-7. The viewfinder is a separate system that uses the top lens. The bottom lens, or taking lens, is either fixed or interchangeable, depending on the camera brand and style (see unit 10).

  VIEW CAMERA. The final camera group is known as the *view camera*, Figure 6-8. These cameras are large and bulky, and most often must be used on a tripod support. They are used to pro-

duce high quality portraits and product photographs. These cameras are available in many shapes and sizes, and are primarily used by professional photographers (see unit 11).

CAMERA CARE

Cameras are precision pieces of equipment. They can actually be called technological instruments

FIGURE 6-7. A typical twin-lens reflex camera. *Mamiya America Corporation*

because of their complex design and construction. A close look at adjustable cameras reveals the many small and precise mechanical parts, Figure 6-9. The electronic system is very complex and extensive. This makes it possible to adjust the camera for many different variables.

Any camera must be properly cared for if good results are to be expected. These points should be followed with all cameras by both amateurs and professionals.

HANDLING. Cameras should be treated with respect. Hold onto a camera with a firm grip so there is no chance of dropping it. Most cameras are designed with a wrist strap or neck strap. Use the strap to ensure that the camera is not dropped. Also, make every effort to keep a camera from striking any solid object while it is being carried. Set the camera where it will not be knocked to the floor.

TEMPERATURE EXTREMES. Cameras and accessories are generally made from quality materials, but extreme heat or cold could cause some permanent damage. Film in the camera may especially be harmed if subjected to high heat. It is best to keep a camera in the same environment as where it will be used.

FIGURE 6-8. View cameras are most often used by professional photographers. *Toyo Division, Mamiya America Corporation*

FIGURE 6-9. Many cameras contain very small mechanical parts, electronic systems, and microcomputers. *Nikon, Inc.*

DUST/DIRT. A camera must be kept free of dust and dirt if quality photographs are expected. Do not use a camera in an environment where a high concentration of dust and dirt are suspended in the air. It is wise to frequently remove dust and dirt from a camera and especially from the lens, Figure 6-10. Compressed air, special lens liquid cleaner, and lens tissue should be used to clean a lens. A lens should never be touched with the fingers as the oils from the skin may damage the lens surface. It is wise to place a skylight or ultraviolet (UV) filter over a camera lens to protect it.

MAINTENANCE. Cameras, like any mechanical and electronic device, need regular maintenance. If the camera uses one or more batteries to power the electronic system, replace them on a regular basis. Usually, this should be done once a year. Also, check all visible moving parts and see if they are operating smoothly. If something appears wrong, it may be wise to have a camera repair expert look at it. Another regular habit to follow is to clean the film chamber. A compressed air system, used carefully, serves this need very well, Figure 6-11. Foreign material within a camera can do nothing but harm.

USE A CASE. Most camera manufacturers will arrange to have cases available for their cameras and lenses, Figure 6-12. A wise photographer will always keep a case on a camera and lens when they are not being used. Most cases for adjustable cameras have a hinged cover that can be removed when the camera is in use. This permits the camera body case to stay in position and provides good protection.

FIGURE 6-11. An environmentally safe propellant must be a critical component of a compressed air system used to remove foreign material from within cameras and other precision photographic equipment. *Falcon Safety Products, Inc.*

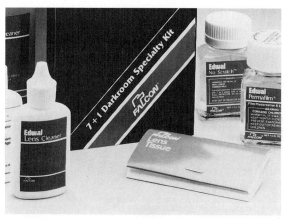

FIGURE 6-10. Correct liquid cleaners and lint-free tissue should always be used to remove dust and dirt from camera lenses. *Falcon Safety Products, Inc.*

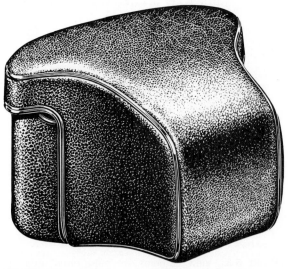

FIGURE 6-12. A leather or plastic case helps protect a camera.

HISTORICAL HIGHLIGHTS

Cameras were designed and understood by scholars long before light-sensitive film became available. The camera obscura was discussed in a knowledgeable way in 1267 by Roger Bacon, an English scientist and philosopher. The term, *camera obscura*, really means a dark chamber or room. Before Bacon's work, Aristotle observed pinhole images as sunlight passed through the gaps between leaves in tree foliage.

No one specific individual can be identified as having invented or developed the camera. Lenses were first used with the camera obscura in the middle 1500s. With this added feature, images were quite clear. Cameras of the moveable type appeared in 1575, but they were large. In fact, they were large enough for people to work inside the darkened chamber or body. Cameras began to be made smaller after light-sensitive materials were developed. They were, though, still much too large for the average person to use, even during the middle 1800s.

It took many years for hand-held cameras to be designed and manufactured. The first Kodak camera was placed on the market in 1888, Figure 6-13. It was loaded with enough flexible roll film for 100 pictures. The price of this camera was $25.00. After all the pictures were exposed, the owner was required to return the entire camera to the factory in Rochester, New York. Here the film was removed from the camera, processed, and prints were made. The camera was reloaded with film, and the camera plus the processed film and prints from the previous roll were returned to the owner. The total cost of the service was $10.00.

Shortly after this innovation in photography, the Eastman Kodak Company marketed the first daylight-loading camera, Figure 6-14. The year was 1891, only 3 years after the first introduction of hand-held cameras to the enthusiastic public. Unlike the camera of 1888, this one could be loaded and unloaded with film by the owner. Also, this film exchange could be done even in a lighted room.

Thousands of these early hand-held cameras were purchased by the eager public. People then,

FIGURE 6-13. The first Kodak camera was made available to the public in 1888. *Eastman Kodak Company*

FIGURE 6-14. This 1891 camera could be loaded and unloaded with film in the daylight by the photographer. *Eastman Kodak Company*

just as now, wanted to capture on film their families, friends, and the scenes surrounding their everyday lives. Amateur photography was born with hand-held cameras in 1888. Today, only a few decades later, cameras are available to nearly everyone.

SUMMARY

Cameras are mechanical-electronic devices that control light. This is necessary so that the exact amount of light reaches the film causing an exposure to be made. It is important to become acquainted with the parts of a basic camera. Once these are known, it is possible to learn how to operate most cameras in a short period of time. In order to maintain a camera's precision, a camera should be handled with care. Exposing a camera to dust and dirt, and rough handling are unacceptable practices. Photographic principles have been known since the 1200s, but easy-to-use cameras have only been available to the public for just over one century.

REVIEW QUESTIONS

Answer these questions to test your knowledge of the unit content.

1. What part of a camera functions similar to the pupil of an eye?
2. Which is the most basic part of a camera?

   A. Aperture   C. Lens
   B. Body   D. Viewfinder

3. Name the part that makes it possible to aim a camera in the correct direction.
4. T/F "Capturing images on light-sensitive film" is the primary purpose of every kind and size of camera.
5. Most entry-level 35mm cameras belong in which category of cameras?

   A. View   C. Single-lens reflex
   B. Twin-lens reflex D. Viewfinder

6. Why can a camera be called a technological instrument?
7. T/F Extreme temperature changes have little effect on the operation of a camera.
8. Camera batteries should be changed:

   A. Every 2 years   C. Once each year
   B. Once each month   D. Every 6 months

9. The camera obscura was first discussed by Roger Bacon (English scientist) in what year?
10. Who first designed and manufactured cameras on a large scale?

PINHOLE CAMERAS

OBJECTIVES   *Upon completion of this unit, you will be able to:*
- *Explain the basic concept of pinhole cameras.*
- *Make a workable pinhole camera.*
- *Take pictures with a pinhole camera.*

KEY TERMS   *The following new terms will be defined in this unit:*

Camera Back	Image Points
Depth of Field	Panchromatic
Film Retainer	Pinhole Camera
Focal Plane	Sheet Film

INTRODUCTION

Images have been observed and recorded using the pinhole principle for many years. Nomadic tribes of North Africa who lived long before the time of Christ observed pinhole images. Countryside scenes were created on the inside of their animal skin tents. The light passed through the small holes in the tent wall where the sun was brightly shining.

Leonardo da Vinci wrote about the pinhole camera principle during the 16th century. Artists used moveable rooms or "camera obscura" to create pinhole images for use in tracing observed scenes. Throughout the years, pinhole photography has been studied and used. Even as late as the early 1940s, the U.S. military made plans to use pinhole cameras.

Today, pinhole cameras are used by the hobbyist and by the photographer interested in conducting experiments. Also, using a pinhole camera is a good way to learn the basics of light and its effect on photographic film and paper.

THE BASIC CONCEPT

The *pinhole camera* is a light-tight enclosure except for the very small hole in one side, Figure 7-1. Light reflected from the scene being photographed is made up of thousands of small light rays or points. Some of these *image points* are reflected through the small

pinhole (aperture). When the image points pass through the aperture, they are reversed. That explains why the image is upside-down and reversed from left-to-right on the *focal plane.*

A pinhole camera must have four basic parts: the light-tight box or chamber, an aperture, a shutter, and a film or paper holder. The chamber of the camera can be one of many shapes and sizes, Figure 7-2. Also, it can be made from various materials such as paper (cardboard), metal, wood, or plastic.

An aperture or pinhole must be positioned at one end or side of the chamber. Usually, the best material to use for this is thin metal foil. The small hole is placed in the foil with a sharp straight pin of the type used by a seamstress. The pinhole aperture on commercial model pinhole cameras are drilled. This makes them more precise and of consistent size.

The shutter of a pinhole camera can be either very simple or elaborate. The simplest shutter is none at all. Actually, it is possible to place a finger over the aperture once the film is loaded into the camera. When an exposure is to be made, the finger can be removed for the exposure time. Replace the finger over the aperture until the camera is in a darkroom where the film will be removed for processing.

A piece of black tape is a convenient shutter. It can be applied easily over the aperture and

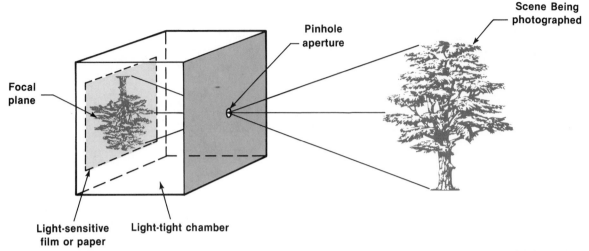

FIGURE 7-1.   The principle of a pinhole camera. It is a light-tight chamber containing a pinhole aperture and light-sensitive film or paper.

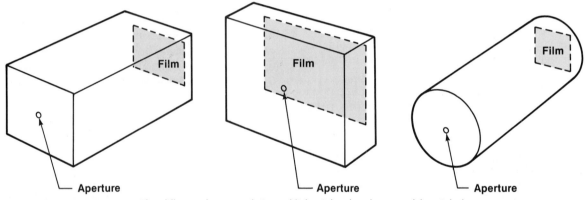

FIGURE 7-2.   The different shapes and sizes of light-tight chambers used for pinhole cameras.

then hinged open when an exposure is being made. An elaborate sliding cardboard shutter is used by some pinhole camera makers, Figure 7-3. It takes more time to design and construct, but it is very convenient to use.

Some type of film holder is needed to make certain the photographic film or paper stays in place during the exposure. Most pinhole cameras use *sheet film,* but some are designed to use roll film in pre-packaged plastic cartridges. The easiest sheet film holder is tape. The film can be positioned on the focal plane and held in place with small pieces of

black plastic tape at each corner. It is possible to design elaborate film holders, but it is unnecessary to do so.

PINHOLE CAMERAS AND PHOTOGRAPHS

Pinhole cameras vary in shape and size as illustrated in Figure 7-2. The criteria for one of these cameras are four basic parts—light-tight chamber, aperture, shutter, and film holder. Anyone making a camera has the choice of how it will look.

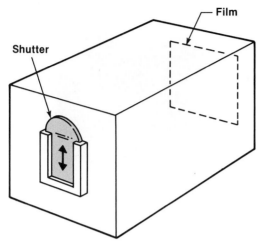

FIGURE 7-3. A sliding cardboard shutter for a pinhole camera.

FIGURE 7-4. A commercial model pinhole camera. Size 126 film cartridges are used in creating color slides, color prints, or black-and-white prints. *David M. Pugh, Time-Field Co.*

It is possible to purchase pinhole cameras from commercial companies. A commercial pinhole camera can be a rather accurate unit, Figure 7-4. The aperture is drilled in a thin piece of brass, thus making the pinhole more accurate than one made by using a straight pin. This camera is made of cardboard, and roll film packaged in commercial cartridges is used. This makes it possible to take color slide film, color print film, or black-and-white print film.

Figure 7-5 shows two black-and-white photographs taken with the camera shown in Figure 7-4. Both photographs show good image detail and

FIGURE 7-5. Two black-and-white photographs taken with the pinhole camera shown in Figure 7-4. The picture on the left clearly shows the statue and the fruit on the tree. In the picture on the right, the 3-inch brass pig was placed on a manhole cover containing one-eighth inch (3.2mm) thick letters. *David M. Pugh, Time-Field Co.*

demonstrate the excellent depth of field that a pinhole aperture is able to provide. *Depth of field* is the image area in acceptable focus from the foreground to the background.

MAKING A PINHOLE CAMERA

Remember, a pinhole camera is only as good as the time and effort that go into making it. Simply stated, take time to design and make a camera that will look good, be easy to use, and take good pictures.

SAFETY CAUTION

Be extra careful when making a pinhole camera. Sharp tools will be used to cut and trim the materials used to form the camera. Be especially careful when using the straight pin to create the aperture.

A commercial pinhole camera made from three pieces of precut cardboard is shown in Figure 7-6. It uses 4 × 5 inch (10.2 × 12.7 cm) sheet film that can be obtained from photographic retail stores. The camera is sold in kit form and contains four items: outer cover, film retainer, camera back, and shutter, Figure 7-7. The camera is assembled as follows:

1. Open the package and carefully remove the four parts.

2. Fold the outer cover containing the aperture on the crease lines and affix the tabs. This holds the cardboard as a three-sided box, Figure 7-8.

3. Fold and secure the film retainer in a similar manner.

4. The camera back is now folded and secured according to the crease lines and precut tabs.

5. Prepare the precut vinyl shutter by removing the adhesive covering. Fold the one end over to form a handle.

6. Adhere the shutter to the aperture opening, Figure 7-9. The aperture area is covered with a clear

FIGURE 7-6. A commercial model pinhole camera. It is designed to use sheet film. *The Pinhole Camera Company*

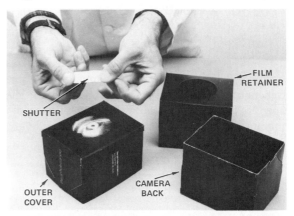

FIGURE 7-7. The four items contained in the pinhole camera kit used to make the camera shown in Figure 7-6. *The Pinhole Camera Company*

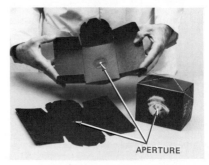

FIGURE 7-8. The outer cover of the pinhole camera shown in Figure 7-6. This part contains the predrilled aperture. *The Pinhole Camera Company*

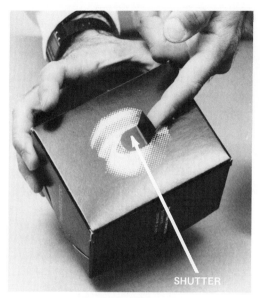

FIGURE 7-9. The shutter correctly positioned and adhered over the aperture. *The Pinhole Camera Company*

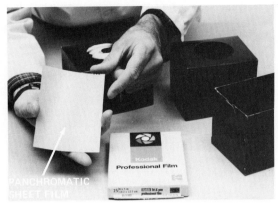

FIGURE 7-10. Sheet film often is used in pinhole cameras. It must be handled in total darkness because it is sensitive to all colors of light. *The Pinhole Camera Company*

plastic tape, thus the adhesive will not pull off the cardboard coating when the shutter is removed to make an exposure.

TAKING PICTURES

The most enjoyable part of working with a pinhole camera is taking some pictures. Excitement of the unknown comes into play because the exact exposure cannot be accurately determined. There is an element of guess that the exposure is correct each time a piece of film is exposed in the camera. Also, a pinhole camera does not have an accurate viewfinder system. This introduces the element of hope when pointing the camera at the selected scene to be photographed.

Good pictures will be obtained if these steps are followed. Patience and attention to details will be strong factors to the success of the pinhole camera photographer.

1. Obtain a 4 × 5 inch (10.2 × 12.7 cm) sheet of black-and-white, *panchromatic* film, Figure 7-10

(see unit 16). Be certain to work with this film in total darkness.

2. Place the film in the *camera back* against the focal plane. The emulsion side of the film must face the aperture. Sheet film contains one or more notches in a corner of the narrow width. When the notches are in the upper-right corner, the emulsion will be facing the photographer.

3. Hold the film in place by inserting the *film retainer* inside the camera back. This holds the film securely while the exposure is being made.

4. Slide the *outer cover* over the film retainer and camera back. This creates a light-tight chamber for the film. The camera can now be brought into a lighted room.

5. Select the scene which is to be captured on film. Set the camera on a solid flat area such as a table. Aim the camera toward the selected scene.

6. Estimate the length of exposure. Use the exposure times in Figure 7-11 as a beginning point.

7. Obtain a wristwatch with a secondhand. It may be wise to ask another person to assist with the exposure times.

8. Remove the shutter and make the exposure, Figure 7-12. Do not move the camera during any part of the exposure process. Camera movement or vibration will cause the photograph to be blurred.

Outdoor Scene Conditions	ISO Film Speed Rating	
	125	400
Bright or hazy sun (Sharp shadows)	6 sec.	2 sec.
Cloudy bright (Soft shadows)	20 sec.	4 sec.
Overcast or open shade (No shadows)	60 sec.	9 sec.

FIGURE 7-11. Suggested exposure times for taking pictures with two different films.

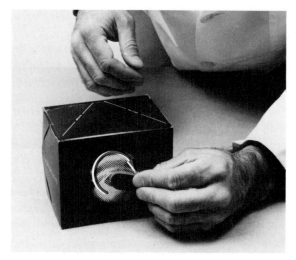

FIGURE 7-12. Removing the shutter so an exposure can be made. *The Pinhole Camera Company*

Replace the shutter when the exposure time is completed.

9. Process the exposed film. Follow the procedures as presented in Chapter 5. Prints from the film negative can be made as presented in Chapter 6.

10. Analyze the results. Make exposure corrections as needed and repeat the process. Seek the best quality possible.

SUMMARY

Pinhole images of scenes were observed for centuries before scientists described their meaning. It was even many years later before pinhole images were recorded on film. The pinhole camera does not contain an actual lens. The light passes through a pinhole-size opening and forms an image on the inside-back of the camera. Surprising picture quality is possible with a pinhole camera. These cameras can be completely hand-made or can be obtained from commercial sources. Taking pictures with a pinhole camera is an exciting experience.

REVIEW QUESTIONS

Answer these questions to test your knowledge of the unit content.

1. Which early scientist wrote about the pinhole camera principle?

    A. Leonardo da Vinci
    B. Joseph Nicéphore Niépce
    C. Aristotle
    D. Roger Bacon

2. T/F Pinhole cameras are limited in their shapes and sizes.

3. The small pinhole in either thin metal or black paper serves as an _____.
4. Name the four basic parts of a pinhole camera.
5. T/F It is essential to have a shutter device on a pinhole camera.
6. T/F Sheet film is used in most pinhole cameras.
7. Black-and-white and _____ film can be used in pinhole cameras.

8. What is the most important requirement when making a pinhole camera?
9. T/F Pinhole cameras can be outfitted with very accurate viewfinders.
10. What serves as a good timing device when making an exposure with a pinhole camera?

─── UNIT 8 ───
VIEWFINDER CAMERAS

OBJECTIVES   *Upon completion of this unit, you will be able to:*
- *Describe the significant features of a viewfinder camera.*
- *Distinguish viewfinder cameras from other types of cameras.*
- *Explain the viewing and lens systems of viewfinder cameras.*
- *Summarize the advantages and disadvantages of using viewfinder cameras.*

KEY TERMS   *The following new terms will be defined in this unit:*

Film Cartridge	*Infinity*	*Parallax Error*
Film Magazine	*LED*	*Rangefinder*
Fixed-focus	*Metering System*	*Snapshots*
Ground Glass		

INTRODUCTION

Of the four camera groups, the viewfinder type of camera is the most popular. There are two main reasons for this fact. First, this camera style or design was the first to be used by manufacturers that produced cameras for public use. As stated in Unit 6, this took place beginning in the late 1880s.

Second, the viewfinder camera is easy to use. It contains few moving parts and has limited adjustments. The photographer need only "aim and shoot," Figure 8-1. Most people are interested in convenience and in seeing the finished print. This fact will continue to make this camera style popular in future years.

THE BASIC CAMERA

Viewfinder cameras vary greatly in the number of moving parts and adjustments, Figure 8-2. The photographer is interested in taking *snapshots* of people, places, and events; thus, the camera must be easy to operate.

Electronics plays an important part in nearly all cameras, but this is particularly true with most viewfinder cameras. Sensitive *metering systems* measure the amount of light falling on the subject being viewed in the viewfinder. Small lights in the viewfinder of some cameras tell the photographer whether the flash is fully charged and ready for use in taking the picture.

FIGURE 8-1. Taking a picture with a viewfinder camera.

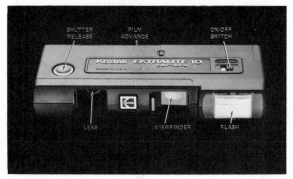

FIGURE 8-2. The parts of a basic viewfinder camera. This camera design uses film packaged in 110 cartridges. *Eastman Kodak Company*

FIGURE 8-3. A compact viewfinder camera that is small enough to be carried in coat pockets and purses. *Chinon America Inc.*

The small lights are often referred to as *LEDs.* These letters stand for *light-emitting diode.* Many viewfinder cameras are equipped with a LED light that blinks when the batteries are low. This is a reminder for the photographer to change the batteries so that the built-in flash and sensing system will continue to operate. On some viewfinder cameras without a built-in flash, a LED light blinks when flash lighting is needed.

CATEGORIES OF VIEWFINDER CAMERAS

Viewfinder cameras are available in several shapes and sizes. They can be grouped according to the size and kind of film used in the camera. Regular light-sensitive film, black-and-white or color, is packaged in either plastic cartridges or thin metal cassettes. Both film packaging methods must be 100% light-tight.

A point to remember is that nearly all cameras are designed to accept one size of film. Also, cameras will accept film that has been packaged a certain way. For example, a 110 viewfinder camera will only accept film packaged in a 110 cartridge. It is impossible to use a film magazine containing a roll of film in the place of a *film cartridge.*

All viewfinder cameras designed to use these prepackaged films are light in weight. A camera without the film cartridge will weigh as little as

3.3 ounces (92.4 g). Because of the light weight and small size, these cameras are often referred to as pocket and compact cameras, Figure 8-3. These cameras are designed to be easily carried in the pockets and purses of the owners. Along with being so easy to use, having a viewfinder camera available at all times makes it possible to capture the "right picture" on film.

VIEWING SYSTEM

The viewfinder camera has a simple viewing system. The system contains two or more pieces of optically *ground glass* that permit the photographer to see the intended picture. The pieces of glass serve as a lens, thus giving the photographer a focused view of the scene to be photographed.

The viewing system on some viewfinder cameras contains an inherent defect. This problem is called *parallax error*. The error is the difference between what the photographer sees in the viewfinder and what the lens captures on film. The rea-

son the error occurs is because the viewfinder and lens are separated from each other.

High-quality viewfinder cameras have been engineered to correct for parallax, Figure 8-4. It is, though, impossible to make a 100% correction. Viewfinder cameras used to take snapshots of family and friends often cause serious frustrations. This is especially true when pictures are taken at the minimum distance for a fixed-focus lens, Figure 8-5. It is very easy to cut off the top portion of the image as seen in the viewfinder. The photographer can correct by tilting the camera slightly upward when taking pictures at close range. The same parallax problem occurs from side-to-side when the viewfinder is positioned on either side of the lens, Figure 8-6. Parallax correction is possible if the photographer remembers to "over aim" the camera when taking pictures up close. No specific guidelines can be given because each camera brand and design is different.

LENS SYSTEM

Most often the lens is nonadjustable, which means it has *fixed focus*. This permits pictures to be taken as close as 4 feet (1.2 m) and up through *infinity*. Some viewfinder cameras contain an extra lens that can be slid over the regular lens. When these two lenses are used together, it is possible to take close-up pictures between 1.5 to 4 feet (.45 to 1.2 m). Further, more elaborate fixed-focus viewfinder cameras have built-in telephoto lenses permitting close-up pictures of subjects far away. See Figure 8-4.

Adjustable lenses are available on some viewfinder cameras, Figure 8-7. This means that it is possible to look through the viewfinder and focus the lens. The lens is adjusted by turning the lens left or right with the fingers. This moves selected lens elements (see unit 13) in or out until the image is focused on the focal plane.

The viewing system and the lens system are tied together in the sophisticated viewfinder cameras. This is done with a *rangefinder* mechanism that determines the distance between the camera and the

FIGURE 8-4. An automatic, compact 35mm viewfinder camera. This camera features a 38mm-76mm zoom lens and built-in multi-mode flash. *Olympus Corporation*

VIEWFINDER SCENE

FINISHED PHOTOGRAPH

FIGURE 8-5. The results when a picture is taken at minimum distance with a "noncorrecting" viewfinder camera and when the viewfinder is directly above the lens.

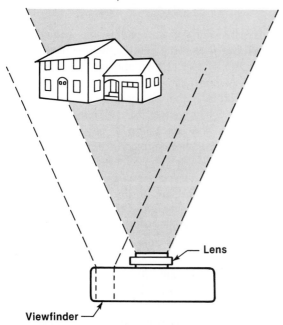

FIGURE 8-6. A top view showing how parallax error occurs from side-to-side (horizontal) when the viewing system is to the left or right of the lens on a viewfinder camera.

subject. The rangefinder system uses a prism, a mirror, the viewfinder lens, and a sight opening, Figure 8-8. This coupled rangefinder-lens system is very accurate in determining the distance between the camera and the subject. Knowing this information often helps in making aperture opening and shutter speed adjustments.

ADVANTAGES AND DISADVANTAGES

Over the years, these cameras have been used to take millions of pictures. People with little to no photographic background can obtain acceptable results. Professional photographers have and continue to benefit by using viewfinder cameras. It is a mistake to think that this camera category is only for amateur photographers interested in taking snapshots.

FIGURE 8-7.   A professional-type, viewfinder camera that includes interchangeable and adjustable lenses, a viewing system with a rangefinder, film speed adjustment, and a dedicated electronic flash system. *Mamiya America Corporation*

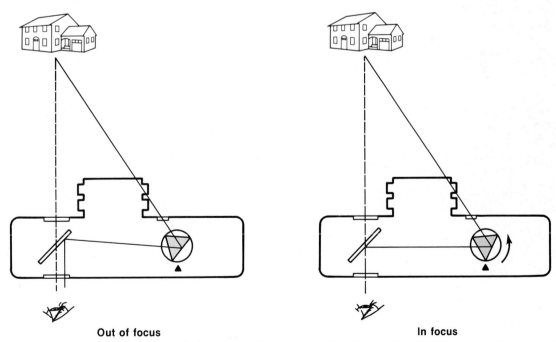

Out of focus   **In focus**

FIGURE 8-8.   Prism movement is used to focus the lens on an adjustable viewfinder camera containing a coupled rangefinder.

ADVANTAGES

- Few moving parts
- Easy viewing in dim light
- Most cameras are small and lightweight
- Quiet shutter operation
- Easy film loading and unloading
- Appropriate cost
- Excellent focusing

DISADVANTAGES

- Limited lens interchangeability
- Viewfinder image generally smaller than actual subject
- Parallax error with most camera designs
- Limited adjustments on some models
- Cartridge film loading not available in 35-mm size

Cameras that produce instant photographic prints are part of the viewfinder camera group. These are special in their operation; thus, the entire content of unit 43 is devoted to instant photography.

FIGURE 8-9. A high quality 35mm rangefinder camera. *Leica Camera Inc.*

SUMMARY

More viewfinder cameras have been manufactured and sold than cameras in the other three categories. People of all ages and abilities find this type of camera useful for general picture taking. Most viewfinder cameras feature easy loading and unloading of film. A minor problem with viewfinder cameras is the parallax error. The camera operator is seldom concerned about this inherent viewing system defect unless close-up pictures are being taken. Finally, the lens is usually nonadjustable on viewfinder cameras. There are, though, some cameras with adjustable and interchangeable lenses, Figure 8-9.

REVIEW QUESTIONS

Answer these questions to test your knowledge of the unit content.

1. List two reasons why the viewfinder camera is the most popular.
2. What do the letters LED mean?
3. T/F Each viewfinder camera will accept a minimum of two and sometimes three sizes of film.
4. T/F Viewfinder cameras are generally very light in weight.
5. T/F The viewing system of a viewfinder camera is often of a complicated design.
6. What is parallax error?
7. T/F Some viewfinder cameras do not have parallax error.
8. Fixed-focus lenses on viewfinder cameras generally permit close-up pictures as close as:

   A. 1′ (.30 m)   C. 3′ (.91 m)
   B. 2′ (.61 m)   D. 4′ (1.2 m)

9. What specific type of viewfinder camera has the viewing and lens systems tied together?
10. T/F Viewfinder cameras are used only by amateur photographers.

SINGLE-LENS REFLEX CAMERAS

OBJECTIVES   *Upon completion of this unit, you will be able to:*
- *Describe the standard SLR camera.*
- *Explain the viewing system of an SLR camera.*
- *Review the many SLR lenses and accessories.*
- *Discuss the advantages and disadvantages of an SLR camera.*

KEY TERMS   *The following new terms will be defined in this unit:*

Bayonet Mount
Film Pressure Plate
Film Speed Dial
Film Sprockets
Flash Hot-shoe

Flash Synch-socket
Focusing Screen
Memo Holder
Microprocessor
Mode Dial

Pentaprism
Rewind Button
SLR
Tripod Socket
Viewfinder Eyepiece

INTRODUCTION

The single-lens reflex (SLR) camera is a precision instrument, Figure 9-1. It is engineered and manufactured for both serious amateur photographers and professional photographers. Images are captured on film under conditions that are within the control of the photographer, a microprocessor, or a combination of the two.

THE STANDARD CAMERA

The name, *single-lens reflex*, is used to identify this camera design because of two factors. First, only one lens is used on the camera at a time. Second, the light reaches the viewfinder eyepiece after it has been reflected from a mirror that moves.

The standard SLR camera contains basic parts and controls that are common to all brands of cameras. From the front and top of a camera, the following parts are visible, Figure 9-2.

BODY.   A light-tight rectangular framework. It is generally made of a combination of metal and plastic.

LENS.   A light gathering system of precisely ground pieces of glass that focus the image on the film. The aperture system is part of the lens system (see unit 14).

FILM ADVANCE LEVER.   A complete movement or pull of this lever advances the film one full frame. While the film is advanced, the shutter system also is being prepared for the next exposure.

FIGURE 9-1.   A single-lens reflex (SLR) camera of the type often used by serious amateur and professional photographers. *Nikon, Inc.*

FIGURE 9-2. The parts and controls from the front and top of a typical 35-mm, single-lens reflex camera. *Minolta Corporation*

FILM REWIND LEVER. This lever is used to return the exposed film back into the light-tight film magazine. It also serves to control the latch holding the camera back and to hold it tightly closed.

SHUTTER RELEASE. Squeezing on this button releases the shutter and exposes the film. The button also serves as a switch to turn on the electronic measuring system within the camera. This system is activated by slightly pressing the button. The electronic measuring system automatically turns off after a short time.

EXPOSURE MODE DIAL. Movement of this dial places the camera in one of several exposure modes. In the programmed or automatic modes, the microcomputer selects the best aperture opening or shutter speed or both. In the manual mode, the photographer can override the microcomputer and select both the aperture opening and shutter speed. The flash mode synchronizes the shutter speed with the electronic flash. Finally, the B position mode permits time exposures. When the shutter release is pressed, the shutter opens and stays open until the button is released. Also, the exposure *mode dial* often contains a position which locks the shutter release. This protects against accidental exposures.

SHUTTER SPEED SELECTOR BUTTONS. These buttons become active when the exposure mode dial is adjusted to the manual position. Pressing the front button (closer to the lens) increases the shutter speed. Pressing the back button decreases the shutter speed.

EXPOSURE COUNTER. Numbers appearing in the small window tell the photographer how many exposures have already been taken on the roll of film.

SELF-TIMER. Moving the lever in a counterclockwise direction readies the timer that will activate the camera shutter. A slight push will start the timer. This permits the photographer about 10 seconds to be part of the picture.

FLASH HOT-SHOE. A bracket designed to hold standard and dedicated flash units on top of the camera (see unit 21).

VIEWING SYSTEM HOUSING. The chamber containing the viewfinder eyepiece and pentaprism.

FLASH SYNCH-SOCKET. A socket that accepts the wire (cord) connection from a flash unit that does not use the electrical contacts in the hot-shoe. Some cameras have "cold shoes," meaning that no electrical contacts are built into the shoe. This requires the flash unit to be attached to the camera with the wire.

FILM-SPEED DIAL. This permits the camera operator to set the ISO film speed rating. This provides important data to the microprocessor in the camera for exposure calculations.

EXPOSURE COMPENSATION DIAL. A special adjustment allowing a photographer to overcome difficult lighting conditions that affect the automatic light-metering system in the camera.

STRAP BRACKETS. These two bracket rings permit a neck strap to be attached. It is wise to place the strap around the neck when using or handling the camera.

The parts and controls are located on all sides of a typical SLR camera. The following impor-

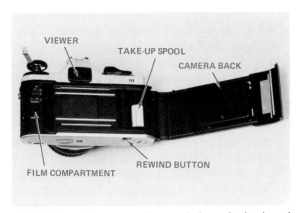

FIGURE 9-3. The parts and controls from the back and bottom of a typical 35-mm, single-lens reflex camera.

tant parts and operator controls are visible from the back and bottom sides of the camera, Figure 9-3.

VIEWFINDER EYEPIECE. This permits the photographer to aim the camera and view the image through the lens prior to making an exposure.

CAMERA BACK. A hinged door that when opened permits film to be loaded and unloaded from the camera. It must be light tight when closed.

FILM INDICATOR. A small window that helps the camera user know if the film is advancing and rewinding properly.

MEMO HOLDER. This is used to hold the end of a film box. It provides the camera operator with valuable information about the film in the camera. The kind, speed, and number of exposures of the film are available at a glance.

FILM CHAMBER. The cylindrical area where the film magazine is located.

SHUTTER. The focal plane shutter that opens and closes to expose the film to the image projected by the lens (see unit 14).

FILM RAILS AND GUIDES. Precision tracks that guide the film through the camera.

FILM SPROCKETS. A shaft containing two pin sprockets that pull the film through the camera. It also measures out the needed film for each new

exposure. The sprockets are activated by the film advance lever.

FILM TAKE-UP SPOOL. A shaft containing a gripping device to hold the end of the film leader. Film is rolled around the spool after each exposure is made. The spool is activated by the film advance lever and is timed with the film sprockets.

FILM PRESSURE PLATE. Under spring tension, the plate presses lightly against the base side of the film. This holds the film firmly against the film rails and between the film guides. This ensures that the film is positioned exactly at the film plane where the lens focuses the image.

CASSETTE PRESSURE SPRING. A piece of spring steel lightly presses against the film cassette. This holds the cassette securely in the film chamber while film is being used in the camera.

GUIDE PINHOLE. A small hole that helps to accurately position a motor film advance system.

MOTOR DRIVE TERMINALS. Electrical contact points that tie together the electrical wiring systems of the camera and motor film advance.

BATTERY CHAMBER COVER. Two small but powerful batteries are placed within this chamber. The energy from these batteries powers the electronic system within the camera.

TRIPOD SOCKET. A threaded chamber in the sturdy base of the camera. The socket permits the camera to be fastened securely to a tripod stand. A motor film advance unit is secured to the camera at this point.

REWIND BUTTON. Pressing this button releases the film sprocket shaft and take-up spool. This permits the film to be returned to the film cassette. It serves to protect against accidental rewinding of the film.

MOTOR DRIVE COUPLER. This cover protects the coupler point for the motor film advance unit.

VIEWING SYSTEM

The significant feature of any single-lens reflex camera is the viewing system. The image seen

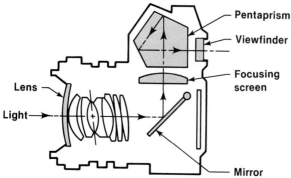

FIGURE 9-4. The five significant elements of the viewing system of an SLR camera.

Besides the lens and viewfinder eyepiece, the viewing system has three main elements, Figure 9-4. These are a hinged mirror, a focusing screen, and a glass *pentaprism*. Penta, the word prefix meaning "five," indicates that the prism simply has five sides. When light enters the prism from the bottom, it reflects from three sides before it can be seen in the viewfinder eyepiece. The prism design corrects the reverse image of the lens and permits the image to be seen "right-reading."

The path of light carrying the image passes through the lens and strikes the mirror, which is set at 45 degrees, Figure 9-5. The mirror is a thin sheet of glass with a front coating of aluminum. It cannot be coated on the back as this would cause double images. When the shutter release is pressed, the mirror rapidly reflexes to an upward position, Figure 9-6. This permits the image-light to reach and expose the film. The mirror also forms a light-tight

through the viewfinder eyepiece is actually coming through the lens. Because of this, the photographer is assured of exposing the film to the same image as seen through the viewing system.

FIGURE 9-5. A cut-away view of an SLR camera which clearly shows the image viewing path through the lens, pentaprism, and viewfinder. *Nikon, Inc.*

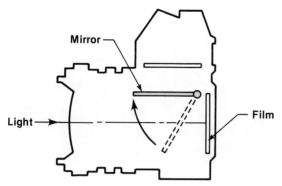

FIGURE 9-6. The mirror moves upward to a horizontal position when the shutter release is pressed. This permits light to reach the film.

seal around the focusing screen so that no light reaches the film through the viewfinder eyepiece.

At 45 degrees, the mirror directs the light to the *focusing screen*, which is located just below the pentaprism. The image is formed on the matt or diffused surface of the screen. From here, the pentaprism reflects the image, focused or nonfocused, to the viewfinder.

Medium to higher priced SLR cameras have interchangeable focusing screens, Figure 9-7. All designs have the common characteristic of a center circle. This circle aids the camera user in focusing the image in the viewfinder (see unit 14). These focusing screens generally are made of high quality plastic. It is possible for the photographer to change

focusing screens on some SLR cameras, Figure 9-8. They must be handled with utmost care. A special tool is used as the focusing screens should never be touched with fingers. Fingerprints cause dust and dirt to accumulate and create difficult viewing of the image.

LENSES AND ACCESSORIES

Being able to view the image through the lens is an excellent advantage of the SLR camera. Adding to this benefit is the fact that many lenses and accessories are available for most brands and models of SLR cameras, Figure 9-9. Lenses of various focal lengths and designs can be quickly attached and removed from the SLR camera lens mount. Most manufacturers use the *bayonet mount*, Figure 9-10, but some of the older cameras still use a threading system. The bayonet mounting method is considered better because it is faster to use and there is less chance of damage to it. The fine threads of the threading method are susceptible to damage in handling and installing in the camera body.

Many accessories are available for SLR cameras. These range from filters (see unit 18) to lens hoods and lens extension tubes (see unit 15). The variety of accessories is almost endless. Today, it is possible for the photographer, amateur or professional, to obtain devices that can make the standard SLR camera more versatile than ever imagined by the originators of this unique camera.

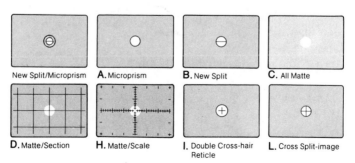

FIGURE 9-7. Several patterns or styles of focusing screens are available for some SLR cameras. *Canon U.S.A., Inc.*

FIGURE 9-8. Changing a focusing screen in an SLR camera. *Pentax Corporation*

FIGURE 9-10. The bayonet mounting system is most often used to fasten lenses to SLR cameras. Each camera manufacturer uses its own style of coupling.

FIGURE 9-9. A wide variety of lenses is available for most brands and models of SLR cameras. *Pentax Corporation*

SLR CAMERA ADVANTAGES AND DISADVANTAGES

Many brands, models, and sizes of SLR cameras are available. A selected few are shown in Figures 9-11 through 9-16. Many others can be obtained from retail and catalog stores throughout many countries of the world.

ADVANTAGES

- A variety of controlled adjustments
- Wide selection of lenses available
- Exposure selections made by a microprocessor
- Wide range of shutter speeds from very slow to very fast
- Through-the-lens viewing system, thus no parallax error
- Through-the-lens light metering system
- Many brands, models, and sizes available

DISADVANTAGES

- Somewhat complex for a beginning amateur photographer
- Shutter system tends to be noisy
- Sometimes bulky to carry and use, especially with the many lenses and accessories
- Costly to repair if this becomes necessary
- Cartridge film loading not available in 35-mm size

SUMMARY

The single-lens reflex camera is the "camera standard" for serious amateur and professional photographers. These cameras have several operator adjustments, making them very flexible and adaptable for many picture-taking situations. The most

FIGURE 9-11. An SLR camera outfitted with a motor film advance system allowing fast action exposures up to 3.5 frames per second. *Minolta Corporation*

FIGURE 9-12. An SLR camera that features the through-the-lens electronic focus control system (TTL-EFC). *Pentax Corporation*

FIGURE 9-13. An interval data recording system can be added to many SLR cameras. With this system, the year, month, day, hour, minute, aperture, shutter speed, or frame number can be imprinted upon a corner of each film frame. *Pentax Corporation*

FIGURE 9-14. SLR camera equipped with accessories permitting rapid exposures with a motor film advance and rewind. A 250-exposure bulk film pack and a powerful flash unit complete with a battery operated system are provided. *Pentax Corporation*

FIGURE 9-15. An ideal medium format SLR camera for the advanced amateur and professional photographer. Fifteen 60 × 45mm negatives are exposed on each roll of 120 size film. *Victor Hasselblad*

FIGURE 9-16. A precision engineered, ergonomically designed 35mm, SLR camera with an autofocus 35mm to 135mm high resolution zoom lens. The mode and setting controls are precisely matched to the fingers while holding the camera. *Olympus Corporation*

favorable advantage of a SLR camera is its viewing system. Being able to look through the lens before taking the picture permits the operator to see the scene before activating the shutter. The wide variety of interchangeable lens is also very favorable. A major disadvantage of this camera style is the cost of repair. The SLR camera is and will continue to be very popular for years to come.

REVIEW QUESTIONS

Answer these questions to test your knowledge of the unit content.

1. Who uses single-lens reflex cameras?
2. Which of the following four selections is often used to activate the light meter on SLR cameras?

   A. Self-timer
   B. Film-speed dial
   C. Shutter release
   D. Film advance lever

3. Which exposure mode setting permits the shutter to be held open?

   A. Automatic   C. Manual
   B. B   D. Flash

4. Camera self-timers allow the photographer about how much time to become part of the picture?

   A. 2 seconds   C. 20 seconds
   B. 30 seconds   D. 10 seconds

5. What is the purpose of the film indicator on a 35-mm SLR camera?
6. T/F The film rewind button on an SLR camera helps to protect against accidental rewinding of the film.
7. How many sides does a pentaprism in an SLR camera viewing system have?

   A. 2   C. 4
   B. 3   D. 5

8. T/F Focusing screens can be interchanged on all SLR cameras.
9. The _____ lens mounting method is used on most SLR cameras.
10. T/F Through-the-lens metering is not often used on SLR cameras.

TWIN-LENS REFLEX CAMERAS

OBJECTIVES *Upon completion of this unit, you will be able to:*
- *Describe a twin-lens reflex camera.*
- *Explain the viewing system of a twin-lens reflex camera.*
- *Identify features of twin-lens reflex cameras.*
- *Discuss advantages and disadvantages of twin-lens reflex cameras.*

KEY TERMS *The following new terms will be defined in this unit:*

Light Meter Acceptor	TLR
Magnifier	Viewing Lens
Medium Format	Viewing System Hood
Taking Lens	

INTRODUCTION

The *twin-lens reflex* (TLR) camera has been used by serious amateur and professional photographers for years, Figure 10-1. The camera is designed with two complete lens systems that are very closely related. The viewing lens gathers the image-light, permitting the photographer to easily aim the camera. The taking lens also gathers the image-light and directs it to the film. The use of "reflex" refers to a mirror being permanently set at 45 degrees, causing the exact image from the viewing lens to be directed to a viewing screen.

THE BASIC CAMERA

Twin-lens reflex cameras contain basic parts and operating controls. To properly operate a camera of this design, it is valuable to know the purpose of each part, Figure 10-2.

BODY. A light-tight rectangular box that serves to protect the film from unwanted light. It is the base or framework of the camera to which all other parts and controls are fastened.

VIEWING LENS. A series of lens elements designed to accurately gather the image-light, allowing the camera user to see the image before the film is exposed.

TAKING LENS. A series of lens elements that gather the image-light. It is focused directly to the film plane for capturing the visible image on the light-sensitive film.

VIEWING SYSTEM HOOD. Serves as a cover for the viewing system when closed down onto the camera body. Also serves to shield light from the viewing glass, permitting the image to be seen in the viewfinder much easier.

APERTURE CONTROL. Permits the photographer to set the desired aperture (lens opening)

FIGURE 10-1. Twin-lens reflex cameras and their many accessories have been a standard for photographers for many years. *Mamiya America Corporation*

according to the lighting and scene conditions (see unit 14).

SHUTTER CONTROL.   Used to adjust the shutter speed based upon the lighting and scene conditions (see unit 14).

SHUTTER RELEASE.   Pressing this button trips the shutter so it will open and close, thus exposing the film to the image.

FOCUSING KNOB.   Turning this knob moves both the viewing and taking lenses closer or farther away from the focal plane. This permits both lens systems to be accurately focused at the same time.

FILM ADVANCE.   A hand-operated crank used to advance the film through the camera. The crank also serves to cock the shutter in preparation for the next exposure.

FLASH CONTACT.   A socket or outlet that accepts the wire (cord) connection from a flash unit.

Often these are referred to as the PC socket and PC cord.

LIGHT METER ACCEPTOR.   Gathers light reflected from the subject and transmits it to the light meter. This permits the camera user to correctly adjust the shutter and aperture.

VIEWING SYSTEM

An important feature of TLR cameras is the separate and versatile viewing system, Figure 10-3. The viewing lens gathers the image-light and projects

FIGURE 10-2.   The basic parts and controls of a twin-lens reflex camera. *Yashica Inc.*

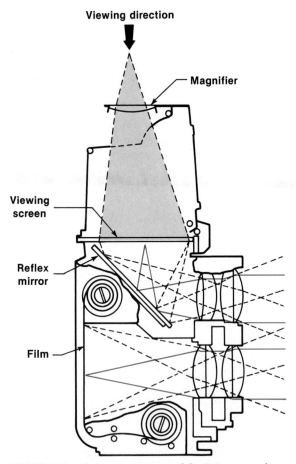

FIGURE 10-3.   The viewing system of the TLR camera showing the standard top viewing position of the photographer.

it to a mirror that is positioned at a 45-degree angle. The mirror diverts the image-light to the viewing screen. This screen is located in a flat position on top of the camera body, making it 90 degrees to the viewing lens. When the photographer looks into the viewing hood, it is possible to see the image on the viewing screen. By turning the focusing knob forward or backward, the image is brought into sharp focus. When this occurs, the image through the taking lens is also brought into focus on the film focal plane. The image is laterally reversed from left-to-right, but this does not cause serious viewing problems.

It is possible to increase the image size on the viewing screen by using the *magnifier* (see Figure 10-3). This special magnifying glass swings out from the viewing system hood directly in the sightline of the photographer. The magnifier makes it possible to adjust the focus of both lenses to a precise setting.

A standard feature of the viewing system hood permits the camera user to quickly aim and capture the image on film, Figure 10-4. The line of sight is directly through a small square opening in the back plate and a larger square opening in the front hood piece, Figure 10-5. Using the viewing

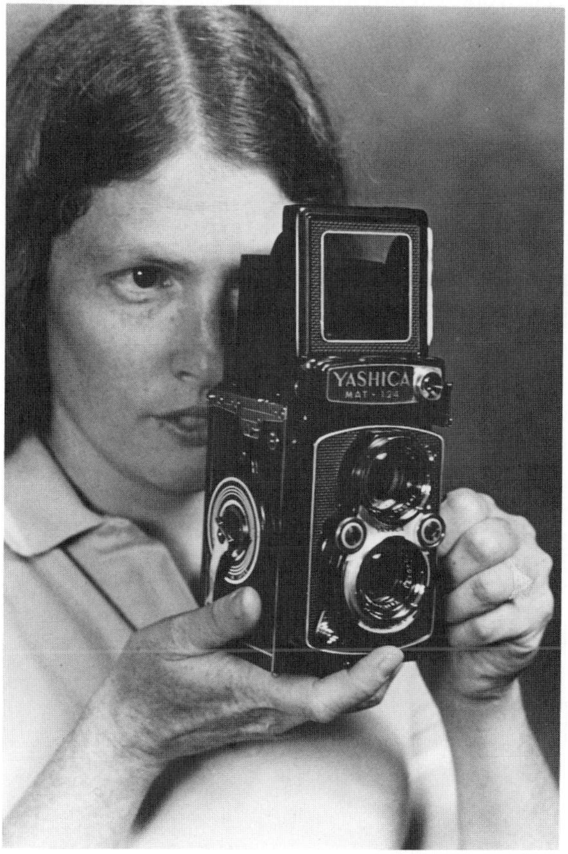

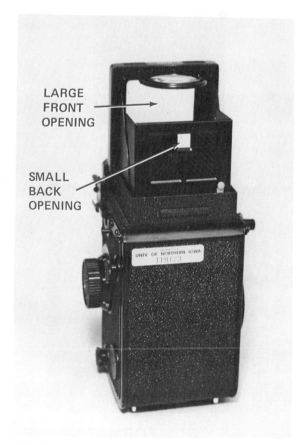

FIGURE 10-4. Using the direct viewing method of aiming the camera.

FIGURE 10-5. The rectangular openings in the front and back pieces of the viewing hood permit quick aiming of the TLR camera.

LARGE FRONT OPENING

SMALL BACK OPENING

system in this manner permits quick aiming of the camera. Often the photographer will take action shots, such as at a sporting event. The standard viewing system hood can be replaced with a special prism and viewfinder eyepiece that makes it possible to use the camera in much the same way as a single-lens reflex camera. Many other accessories are available for TLR cameras, making this camera grouping very flexible and unique.

35 MM FORMAT

6 X 6 CM FORMAT

FIGURE 10-6. A comparison of the 35-mm camera and film format and the 6 × 6 cm format of the twin-lens reflex camera.

USING THE TLR CAMERA

The twin-lens reflex camera is designed to produce images in the *medium-format* range. This means that the film frame size is larger than the 35-mm single-lens reflex camera but smaller than the smallest view camera. Photographers often like the TLR camera because of its large, square, $2\frac{1}{4}'' \times 2\frac{1}{4}''$ (6 × 6 cm) film format. Because the image format is square and large, it has about four times the amount of image area as the 35-mm image format, Figure 10-6. The larger film size, 120 for 12 exposures and 220 for 24 exposures, generally permits photographic enlargements of higher quality.

Twins-lens reflex cameras are used for portrait work, landscapes, and weddings, Figure 10-7. The built-in light metering system permits accurate exposures on a repetitive basis. Interchangeable lenses on some models give the photographer a great deal of flexibility.

TLR CAMERA ADVANTAGES AND DISADVANTAGES

Before single-lens reflex cameras became popular, twin-lens reflex cameras were used frequently by serious amateur and professional photographers. These were used due to their "portability" as compared to the bulky view cameras. Today, TLR

FIGURE 10-7. A photographer using a twin-lens reflex camera to take a picture.

cameras are being used by the same type of people as in past years and for the same reasons. The square film format and the large film size make TLR cameras useful for a variety of photo-taking needs.

ADVANTAGES

- Large film size, thus excellent enlargements are obtained
- Convenient through-the-lens focusing due to the viewing and taking lenses being tied together
- Critical focus magnifier available to give that "extra" image detail when needed
- Image always visible in viewfinder, even when the shutter is activated
- Convenience of both waist-level and eye-level "shooting"

DISADVANTAGES

- Some parallax error, although current TLR cameras have been corrected for this problem
- Image laterally reversed from left-to-right in the waist-level viewfinder
- Camera somewhat large and bulky compared to viewfinder and SLR cameras
- Many models do not have interchangeable lens

SUMMARY

Twin-lens reflex cameras contain two lens systems. One is part of the viewing system, and the other is used to take the pictures. A mirror at a 45-degree angle reflects the image from the viewing lens to the ground viewing glass. This shows the photographer the image that will be captured on film. Parallax error is a slight problem with some TLR cameras. The square picture format is considered a useful feature of this basic camera category.

REVIEW QUESTIONS

Answer these questions to test your knowledge of the unit content.

1. Why is the twin-lens reflex camera called a "reflex" camera?
2. T/F The TLR camera case is generally rectangular in shape.
3. Which camera control moves both of the TLR camera lenses closer or farther away from the camera?
4. The TLR camera viewing screen is positioned at what angle to the viewing lens?

   A. 90°   C. 45°
   B. 180°   D. 30°

5. T/F The image seen on the ground glass viewing screen of a TLR camera is laterally reversed from left-to-right.
6. What is used to increase the image size on the viewing glass of a TLR camera?
7. T/F The photographer must always look down into the viewing screen of a TLR camera to see the image.
8. About how much larger is the picture image area of a TLR camera than that of a 35-mm SLR camera?

   A. 8 times   C. 4 times
   B. 10 times   D. 6 times

9. TLR cameras produce images (pictures) in a _____ format.
10. T/F Parallax error is a serious problem with TLR cameras.

VIEW CAMERAS

OBJECTIVES *Upon completion of this unit, you will be able to:*
- *Identify the parts of a view camera and its viewing system.*
- *Describe the design of view cameras.*
- *Take basic pictures with a view camera.*
- *Explain the advantages and disadvantages of a view camera.*

KEY TERMS *The following new terms will be defined in this unit:*

Camera Tilt	*Lateral Shift*	*Rise and Fall*
Distortion Control	*Perspective Control*	*Swing Control*
Film Holder	*Reflex Viewer*	*Viewing Glass*

INTRODUCTION

The view camera is one of the four basic camera groups, Figure 11-1. It is considered to be a "large format" camera because of the large film size. The

FIGURE 11-1. A view camera is large and bulky, but its flexibility permits photographers to obtain high quality photographs. *Toyo Division, Mamiya America Corporation*

sheets most often range from between 4″ × 5″ (10.2 × 12.7 cm) and 8″ × 10″ (20.3 × 25.4 cm) in size. Roll film is used on some camera models. View cameras are used by professional photographers because it is possible to achieve results that are impossible with other camera types.

CAMERA PARTS AND VIEWING SYSTEM

View cameras are precise instruments, but unlike cameras in the other groups, they have many of their parts exposed to the user. Study Figure 11-2 to become acquainted with the parts and adjustments of a view camera. All of the parts and adjusting devices are easily accessible to the user.

The viewing system of a view camera is simple and direct, Figure 11-3. Without film in the camera, light from the lens shines directly on the *viewing glass*, Figure 11-4. This causes the image to be upside-down and reversed left-to-right. The viewing glass has been etched, causing it to have a diffused or milky appearance. Because of this special surface, the image from the lens is visible. Sometimes the viewing glass is prepared by sandblasting, thus the term "ground" glass is used.

The image on the viewing glass is often faint due to the room light. To darken the area around

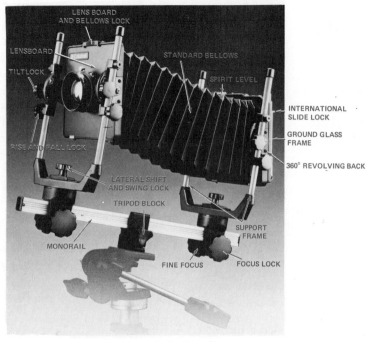

FIGURE 11-2. A standard 4″ × 5″ view camera with its many parts identified. *Calumet Photographic Inc.*

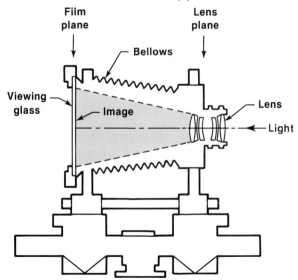

FIGURE 11-3. The direct viewing system of the "large format" view camera. The bellows permits movement and maintains a darkened chamber between the lens plane and film plane.

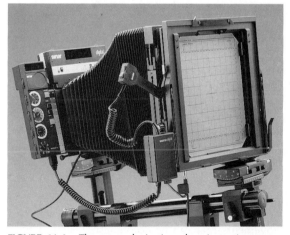

FIGURE 11-4. The ground viewing glass is an important part of a view camera. Lines etched in the "frosted" glass assist the photographer in positioning the image. *Sinar Bron, Inc.*

FIGURE 11-5. A reflex viewing hood eliminates the need for a focusing cloth. *Sinar Bron, Inc.*

the viewing glass, a black cloth, called a focusing cloth, is clipped to the camera back. This makes it possible to easily see the projected image. A special attachment called a *reflex viewer* can be fastened to the viewing glass, Figure 11-5. This eliminates the need for a focusing cloth and makes the image right-side-up but still reversed left-to-right.

VIEW CAMERA DESIGN

View cameras are considered by many people to be large and bulky. Generally speaking, that is true because the use of the camera demands that it have a number of moving parts that no other camera has. Photographers who need the flexibility offered by a view camera are not concerned about its large size and inconvenience or its bulkiness.

View cameras offer the photographer the opportunity to control *distortion* and *perspective* in the finished photograph. There are four standard movements that can be accomplished with a fully adjustable view camera, Figure 11-6. The *tilt* of the camera is adjusted when the lens board and the film holder are rotated from top to bottom from the

vertical position. The *swing control* is adjusted when the film board and the lens board are swung from right to left. The *rise and fall* adjustment is controlled by raising and lowering both the lens board and the film holder. The *lateral shift* is made by moving the lens board and the film holder from side to side. The long bellows between the lens board and the film holder permit these four basic adjustments to be made. This makes view cameras very versatile photographic instruments.

Originally, view cameras were made almost entirely of wood. Some metal was used for the brackets and controls, but the framework was all wood. Today, it is possible to purchase view cameras made of both natural and synthetic materials. View cameras are almost ageless. A view camera manufactured decades ago can be used to take high-quality photographs today, because parts that do become worn can be replaced easily.

USING A VIEW CAMERA

Because view cameras are large and bulky, it does take time and effort to prepare one for use. One of the first steps in using a camera of this type is to select the lens to serve the intended purpose. A wide variety of lenses is available for view cameras, and each lens is capable of achieving certain photographic needs. It takes a considerable amount of experience to know which lens to use at the appropriate time.

Because view cameras are heavy and bulky, it is important to fasten them to a heavy-duty tripod or stand, Figure 11-7. Next, attach a cable release to the shutter control which is found on the front part of the lens. The cable release permits the photographer to trip the shutter without touching the camera and causing the camera to vibrate.

The next step is to aim and focus the camera. This step can take a considerable amount of time depending on the object or subject that is being photographed. The photographer must look at the image that is appearing on the viewing glass. The viewing glass is located precisely in the same

SIDE VIEW

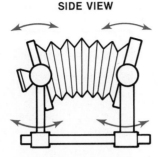

TILT: Rotating (tilting) the front
or back of the camera forward
or backward.

TOP VIEW

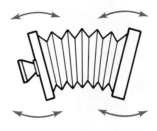

SWING: Rotating (swinging)
the front or back of the camera
left or right.

SIDE VIEW

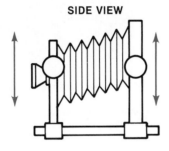

RISE AND FALL: Vertically
raising or lowering the front
or back of the camera.

TOP VIEW

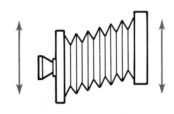

LATERAL SHIFT: Moving the
front or back of the camera
laterally to the left or right.

FIGURE 11-6.   The four standard movements which make the view camera truly "adjustable." *Calumet Photographic Inc.*

position where the film will be located once it is placed into the camera.

Several factors can be controlled by carefully adjusting the camera according to its four basic movements. These special control features include vertical perspective, horizontal perspective, vertical image placement, horizontal image placement, and depth of field. Depth of field is described in unit 13.

At this point, the photographer must load the sheet film into a *film holder*, Figure 11-8. This must be done in total darkness. Most film holders will permit two sheets of film to be loaded at one time. The film is protected from being exposed by a flat plate (dark slide) that slides in front of each

piece of film. Once the film holder is placed in the camera, the protective slide facing the lens can be pulled from the film holder. This allows the light that passes through the lens to reach the film.

The photographer is now ready to take the exposure. A light meter is used to determine the proper aperture and shutter settings (see unit 14). Without causing the camera to vibrate in any way, the cable release is pushed to activate the shutter. This will permit the light to pass through the lens and cause the image on the film to be exposed. If a second exposure of this scene is desired, the film slide must be slid back into place to protect the film from unwanted light. The film holder is removed from the camera and turned around so that

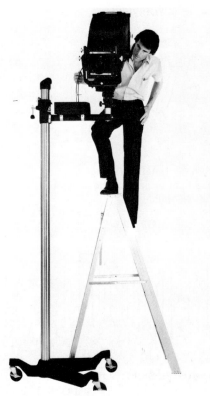

FIGURE 11-7. A heavy-duty stand is needed to hold a view camera steady. *Omega Arkay*

FIGURE 11-8. A view camera film holder used to hold two sheets of film. *Lisco Products Co.*

the second sheet of film is in the proper position. The procedure as just described then would be repeated for a second exposure. The exposed film then is ready for processing (see chapter 5).

VIEW CAMERA ADVANTAGES AND DISADVANTAGES

View cameras are important pieces of equipment, especially to professional photographers. Several brands and models are available that provide the photographer with a choice, Figure 11-9. The camera selection depends greatly on the needed end results.

Some view camera designs permit special set-ups. Using a special adapter, it is possible to

FIGURE 11-9. Several accessories are often available to provide even greater flexibility for photographers who use view cameras. *Toyo Division, Mamiya America Corporation*

FIGURE 11-10. A 35-mm SLR camera can be fastened to some 4″ × 5″ view cameras. *Anthony F. Esposito, Jr.*

convert a 4″ × 5″ view camera into a 35-mm view camera, Figure 11-10. A photographer prepared to work with 35-mm film will find a converted camera like this very useful. The results of using the front tilt feature are shown, Figure 11-11. Using the 35-mm view camera without the front tilt feature gives out-of-focus results, Figure 11-12. Whenever image clarity at close range is needed, view cameras should be used.

ADVANTAGES

- Large film sizes
- Distortion and perspective control
- Variety of special-use attachments

FIGURE 11-11. A sharp focus image results when the front tilt feature of the 35-mm view camera is used. *Anthony F. Esposito, Jr.*

FIGURE 11-12. An out-of-focus image (*top half*) resulted when the front tilt feature of the camera was not used. *Anthony F. Esposito, Jr.*

- Direct viewing of the image through-the-lens
- Excellent photographic print enlargements
- Wide selection of lenses

DISADVANTAGES

- Heavy and bulky
- Requires use of tripod or stand
- Image difficult to see on viewing glass
- Takes considerable time to set up and use
- Image lost on viewing glass after film holder is inserted
- Photographer skills needed to know when and how to use the four standard movements

SUMMARY

Professional photographers often use view cameras to achieve results that are difficult to impossible with other types of cameras. The four basic camera movements are tilt, swing, rise and fall, and lateral shift. These movements make it possible to position the lens and film plane separate from each other. By doing this, the images are formed perfectly on the film. This gives quality photographs for special publication and advertising photography.

REVIEW QUESTIONS

Answer these questions to test your knowledge of the unit content.

1. View cameras are considered to be _____ _____ cameras.
2. T/F Professional photographers use view cameras more often than do amateur photographers.
3. Which part of the view camera is used to hold the lens in place?
4. T/F The image is reversed left-to-right on the viewing glass of view cameras.
5. A reflex viewer on a view camera eliminates the need for which of the following?

   A. Viewfinder   C. Focusing cloth
   B. Viewing screen   D. Film

6. List the four standard movements that can be made on a fully adjustable view camera.
7. Why should a view camera always be placed on a heavy-duty tripod when the photographer is taking pictures?
8. How many sheets of film can be loaded into a standard film holder?

   A. 2   C. 4
   B. 6   D. 3

9. What parts of a film holder keep the film from being exposed when it is not in the camera?
10. T/F The bellows on a view camera maintains a darkened chamber between the lens plane and film plane.

CHAPTER THREE

CAPTURING THE SCENE

FUNDAMENTALS OF PICTURE TAKING

OBJECTIVES   *Upon completion of this unit, you will be able to:*
- *Practice good safety habits while taking pictures.*
- *Identify and make good use of photographic equipment used to complement cameras.*
- *Load and unload film from a camera with accuracy and confidence.*
- *Hold a camera for precision horizontal and vertical pictures.*

KEY TERMS   *The following new terms will be defined in this unit:*

Cable Release	Tripod
Film Protective Bag	Unipod
Neck Strap	X-ray Exposure

INTRODUCTION

Picture taking is exciting. It is both fun and relaxing for the amateur photographer. To the professional photographer, taking pictures is considered work, but there is still a level of excitement whenever a camera is being used. The content in this unit and entire chapter provides helpful information for capturing the selected scenes on photographic film.

PRACTICING SAFETY

Anyone using a camera must observe safe habits. Cameras in themselves are not dangerous, but the photographer's selection of location while taking pictures can become a serious problem. Make a point of observing the following safe picture taking habits. These pointers are not inclusive, but they do provide a good reference for photographers of all abilities.
- Respect heights. Make certain that a substantial ladder is used when there is need to be higher than the floor or ground, Figure 12-1. It is best to have someone hold the ladder secure while the photographer is busy taking pictures.
- Beware of cliffs and dropoffs. Taking pictures in mountainous areas can be dangerous. Watch where cliffs and sudden

FIGURE 12-1. A photographer should always use a sturdy ladder when there is need to be higher than the ground or floor. Also, it is wise to ask someone to hold the ladder to prevent it from tipping and falling over.

76

dropoffs occur. Backing off a cliff could be fatal.

- Stay clear of street and road traffic. Photographers do not mix well with cars, trucks, and motorcycles. If it is necessary to stand in a street or road to take a picture, be certain that motor vehicles are not present. Ask someone to observe and to inform if a vehicle is approaching.
- Caution is advised when taking pictures near a body of water, such as a river, lake, or ocean. Good swimmers have been lost because they fell into water unexpectedly, Figure 12-2. A photographer wanting that "just right" location should always consider whether it is a safe choice.

- Consideration for people who happen to be in the area when pictures are being taken is wise for any photographer. Their safety and the subject's safety must always be considered by the person holding the camera.

USEFUL EQUIPMENT

There are many pieces of equipment designed to make photography interesting and easy to accomplish. Besides an adjustable 35-mm camera with a standard 50-mm lens, it is useful to have one or two additional lenses. A wide-angle lens and a telephoto lens are helpful in obtaining the desired scene (see unit 13).

A *tripod* is a three-legged stand that holds a camera in a specific and steady position, Figure 12-3. Tripods are available in many different sizes

FIGURE 12-2. Water, boating, and photography—all add up to enjoyment. Always stay seated while taking pictures in a small boat. *Alan Jackson*

FIGURE 12-3. A tripod is useful in holding a camera in a specific location. *The Saunders Group*

and strengths. It is important to select one that is designed to fit the picture taking needs of the photographer. Single leg stands called *unipods* are useful for holding the camera steady when a standard tripod is not available. A unipod consumes less space in a photographer's supply pack than a tripod does.

A photographer also will find a cable release, a flash unit, and some filters very useful. A *cable release* is fastened to the camera shutter release, and when the pin is pushed, it releases the shutter, Figure 12-4. Flash units (see unit 21) and filters (see unit 17) add to the scope and flexibility of taking pictures.

FIGURE 12-4. A cable release is used to activate the shutter so that the camera is not moved while taking a picture.

CAMERA AND EQUIPMENT PROTECTION

Photographic equipment must be treated with care. Precision mechanical parts, electronic circuits, and optical lens glass are used in making cameras, lenses, flash units, filters, and other photographic equipment. Proper handling is critical to its operation, whether it be expensive or inexpensive equipment.

One of the most important protection devices is a *neck strap*, Figure 12-5. It permits the photographer to have the camera ready for use and at the same time keeps it from being dropped and damaged. Neck straps of several widths and lengths have been designed and are available from photographic supply stores.

Photographic equipment bags are very useful for carrying equipment and supplies, Figure 12-6. Many of the bags contain padding, thus considerable protection is provided against damaging blows to the bag. It is wise to purchase an equipment bag that is large enough to hold all needed equipment and supplies for most picture taking.

Cameras, and especially film, should be protected from extreme heat and cold. High heat will cause film to lose its physical and light-sensitive properties. Cameras loaded with film should never be left in the direct rays of a hot summer sun.

FIGURE 12-5. A neck strap is very important to the use and protection of a precision camera. *Falcon Safety Products, Inc.*

Film and other photographic products are best preserved in cool conditions, such as a refrigerator. High-speed films of ISO 1000 and above may show excessive grain if they are refrigerated. Consult the instructions packed with the roll or cartridge of film. Extreme cold, though, reduces the light sensitivity

FIGURE 12-6. Photographic bags are very useful for carrying and protecting camera equipment and supplies. *Pentax Corporation*

FIGURE 12-7. Lead-lined bags should be used to protect photographic film from airport X-ray equipment. *Sima Products Corporation*

of photographic products. Also, moisture condenses on a cool surface when it begins to warm up. This is detrimental to the sensitivity of photographic film and paper. Film should be allowed a few hours to reach room temperature after it is removed from the refrigerator. The film should stay in the sealed package so moisture does not condense.

Keep cameras dry. An extremely cold camera can be placed in a plastic bag when brought into a room temperature area. This will cause the moisture to condense on the bag and not on the camera. If a camera must be used in the rain, cover it with plastic to keep water from reaching it. A photographer's assistant is useful in holding an umbrella over a camera while it is being used in the rain.

Airport X-ray equipment will sometimes expose film whether it is loaded in a camera or kept in the original factory packaging. One way to protect film from unwanted *X-ray exposure* is to use a special *protective bag*, Figure 12-7. These bags are lined with a laminated lead coating and stop most X-rays from penetrating. When photographic equipment and supplies are taken through an airport passenger X-ray check station, they should be placed in a sealed protective bag. This includes packaged film that is both nonexposed and exposed, and cameras containing film. High-speed films are very sensitive to X-rays. The safest procedure is to request hand inspection of all photographic film products.

LOADING FILM INTO A 35-mm CAMERA

One of the most important picture-taking tasks is to properly load and unload film from the camera. Manufacturers of film and cameras have devoted considerable time and effort to making this aspect of photography as easy as possible. The cartridge has eliminated film loading and unloading problems for people using viewfinder "snapshot" cameras, Figure 12-8. Film is placed into and removed from a camera in seconds with virtually no possible way to make an error and expose the film to unwanted light.

Adjustable cameras that use roll film packaged in magazines or cassettes and regular spool rolls must be loaded and unloaded with care.

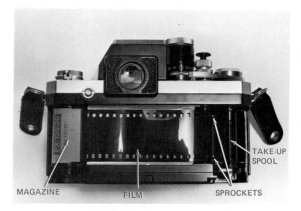

FIGURE 12-10.   Film advanced to the point of the sprockets engaging the film perforations.

FIGURE 12-8.   Viewfinder cameras have been designed for convenient loading and unloading of film for use by the amateur photographer. *Eastman Kodak Company*

FIGURE 12-9.   The camera back opened in preparation for loading film.

Following proper procedure in loading and unloading film in a 35-mm camera is important for consistent results.

1. Obtain a roll of 35-mm film suitable for the pictures that will be taken.

2. Set the exposure mode dial to the flash shutter speed setting.

3. Remove the protective case and place the camera on a firm surface with the lens facing down.

4. Pull the camera back release pin until the back pops open a slight amount.

5. Open the camera back. It is wise to place an object such as the camera case under the back so the hinges are not damaged, Figure 12-9.

6. Place the film magazine (cassette) in the film chamber. **Caution:** It is best to load film in subdued lighting. Be sure the protruding core is pointing in the correct direction. Push the camera back release pin and rewind knob down to engage the film magazine.

7. Pull the film leader across the film guide rails until reaching the take-up spool. Thread the film leader (narrow portion) into the spool.

8. Advance the film by alternately operating the wind lever and depressing the shutter release button. This only needs to be done once or twice until both the top and bottom sprockets engage the film perforations, Figure 12-10.

9. Close the camera back. Make certain it latches by pressing it firmly so the release pin and rewind knob hold the camera back securely. An improperly closed or damaged camera back will permit light leaks, thus exposing the film to unwanted light.

10. Advance the film two complete frames. This pulls film from the magazine that has not been struck by light. The exposure counter should read "1", at this point. While advancing the film, watch the rewind knob. It should turn as film is unwound from the magazine, indicating that the film has been correctly loaded in the camera. If no movement is seen, advance the film one more frame. If there is still no movement in the rewind knob, open the camera and rethread the film onto the take-up spool.

11. Reset the exposure mode dial to the desired setting.

12. Set the film speed dial to match the ISO speed of the film. This is usually done by slightly lifting the dial ring and turning it until the correct number is aligned with a red or orange mark.

13. Place the top of the film box in the memo holder or tape it on the back of the protective case. This provides a ready reference to the kind and speed of film in the camera.

14. Replace the protective camera case. The camera is now ready to be used to take pictures.

UNLOADING FILM FROM A 35-mm CAMERA

Just as with loading the 35-mm camera, it is very important to carefully unload the camera too. For example, the camera back should never be opened in the light with the film extended from the magazine. When the last exposure has been taken, the following procedure should be used.

1. Remove the case from the camera.

2. Raise the rewind crank in preparation for winding the film back into the magazine.

3. Depress and hold the film rewind button. Turn the rewind crank according to the arrow to return film to the magazine, Figure 12-11.

4. Continue rewinding until all of the film, including the leader, is inside the magazine.

5. As when loading the film, open the camera back in subdued lighting and remove the film magazine.

6. Close the camera back, and position the camera so it does not fall from the work location.

7. Place the exposed film in a protective container. It is now ready for processing.

8. Reload the camera with a new roll of film.

USING A CAMERA

Cameras are made to be used; thus, make it a point to have one available at all times. Also, have the camera loaded with film to fit the occasion, and have it preset to the most likely picture-taking situation. Many photographers will carry two cameras loaded with film. One camera will have black-and-white film, while the other will have color film. This provides needed flexibility.

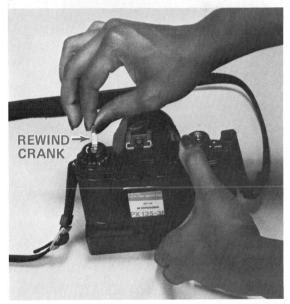

FIGURE 12-11. Rewinding exposed film back into the film magazine.

Taking good pictures is a creative art. A photographer with a camera is similar to a painter with brush and paints. Both are artists who must see the scene and then capture it either on film or on canvas.

Holding a camera correctly helps to capture the desired scene. Always hold the camera firmly so it will not slip from the fingers when the shutter release is squeezed.

Cameras are designed with the shutter release on the right side. This forces the photographer to use the right index finger to operate the shutter release. Sometimes this is inconvenient for a left-handed person.

Figures 12-12 through 12-15 show examples of how a camera can be hand held. Most pictures

FIGURE 12-13. The best positions for the camera and hands when taking vertical pictures.

FIGURE 12-12. The standard hand positions for taking horizontal pictures. Always keep arms close to the body.

FIGURE 12-14. Take advantage of walls and other solid objects to hold the camera steady.

FIGURE 12-15. A prone position with elbows firmly planted helps to keep the camera from moving.

are taken when cameras are held by the photographer. This is appropriate if shutter speeds are 1/60 of a second or faster (see unit 14). It is always wise to use a tripod for slow shutter speeds below 1/60 and when using long (telephoto) lenses.

SUMMARY

Safety is an attitude. After the proper attitude is established, safe practices then will become natural. Along with human safety, it is valuable to protect photographic equipment and supplies with utmost respect. A wise photographer is one who thinks, plans, and follows reasonable behavior. Consider the rights and privileges of others. Honor people's wishes when taking their picture. In short, follow good "photo manners."

REVIEW QUESTIONS

Answer these questions to test your knowledge of the unit content.

1. Why is picture taking fun?
2. Identify five safety precautions that a photographer should observe when taking pictures
3. Which of the following camera lenses would be considered a luxury for the average photographer?

   A. Telephoto   C. 50 mm
   B. Fisheye   D. Wide angle

4. T/F Tripods and unipods are designed to serve the same general purpose.
5. A photographer should always use a _____ so that the camera is not dropped and damaged during photo sessions.
6. T/F X-rays such as those used in checking luggage and personal belongings in airports should not be a worry to the photographer.
7. It is best to load and unload film from a 35-mm camera in _____ lighting.
8. Why should the camera film speed dial and the ISO speed rating of the film match?
9. Before removing an exposed roll of film from a 35-mm camera, it is critical to do the following step:

   A. Rewind the film into the magazine
   B. Open the camera back
   C. Remove the case strap
   D. Advance the film beyond the last exposure

10. The slowest shutter speed that a photographer should attempt to hand-hold a camera is _____.

SELECTING AND USING LENSES

OBJECTIVES   *Upon completion of this unit, you will be able to:*
- *Name the three lens surfaces and the six lens types.*
- *Describe lens focal length and angle of view.*
- *Focus all types of lenses and utilize depth of field to good advantage.*
- *Choose the best lens accessory for a given situation.*

KEY TERMS   *The following new terms will be defined in this unit:*

Angle of View	*Fisheye Lens*	*Multiple Element Lenses*
Auxiliary Close-up Lens	*Flat (lens surface)*	*Nodal Point*
Bellows	*Focal Length*	*Rear Nodal Point*
Concave (lens surface)	*Focal Plane*	*Refraction*
Converging Lenses	*Focusing Screen*	*Split-image Focusing*
Convex (lens surface)	*Hyperfocal Distance*	*Supplementary Lens*
Depth of Field	*Lens Aberration*	*Teleconverter*
Diverging Lenses	*Lens Element*	*Zoom Lens*
Extension Tube	*Lens Group*	

INTRODUCTION

The lens is a critical component of any camera. To obtain sharp photographs, a quality lens must be used. Many kinds, sizes, and styles of lenses are available for adjustable cameras with interchangeable lenses, Figure 13-1. The photographer must select a suitable lens to capture a given scene in the desired manner.

THE BASIC LENS

A lens is designed to gather and bend light, see Figure 13-5. After bending, the light is directed to a specific point called the *focal plane*. This is where the image is in focus and where the film is located. The bending of light is called *refraction*; thus, light is refracted through a lens.

Glass used in camera lenses must be manufactured and shaped very accurately. Only three basic surface shapes can be used to prepare a lens to refract light. These surfaces are *flat, convex,* and *concave,* Figure 13-2.

FIGURE 13-1.   A wide selection of lenses is available for adjustable cameras with lens interchangeability. *Tamron Industries, Inc.*

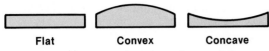

Flat   **Convex**   **Concave**

FIGURE 13-2. The three basic lens surfaces.

With the three basic surfaces, the six lens types are formed, Figure 13-3. Three of the lenses—plano convex, double convex, and meniscus convex—are *converging* or plus lenses. Refracted light from these lenses is directed to the selected focal plane. The *diverging* or minus lenses cause the refracted light to spread.

MULTIPLE ELEMENT LENSES

Only very simple cameras have been designed to use lenses made of one piece of glass. Lenses for all other cameras are made up of several individual pieces of glass, Figure 13-4. Each piece of glass is a lens in itself and refracts light. When several lenses are used together to form a given camera lens, each individual lens is called an *element*. Lenses for 35-mm SLR cameras can have as few as three elements and as many as 16 elements.

The elements making up a lens are positioned in precise locations within the barrel. Some elements are cemented together, while others are

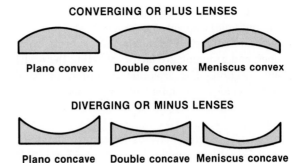

CONVERGING OR PLUS LENSES

Plano convex   **Double convex**   **Meniscus convex**

DIVERGING OR MINUS LENSES

Plano concave   **Double concave**   **Meniscus concave**

FIGURE 13-3. The six basic types of lenses.

positioned by themselves. This is called grouping of lens elements, thus a camera lens may contain seven elements and six *groups*. This means that two elements have been cemented together and five elements are positioned separately.

FOCAL LENGTH

The *focal length* (FL) of a lens is important for a photographer to know. It is defined as "the distance from the optical center of the lens to the focal plane when the lens is focused at infinity," Figure 13-5. With a single element lens, it is easy to determine the center, called *nodal point*, because it is the exact center of the glass.

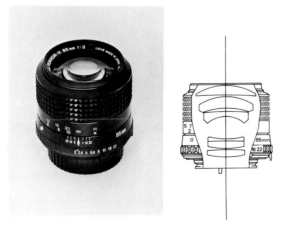

FIGURE 13-4. Several lens elements are used together to form a quality camera lens. *Minolta Corporation*

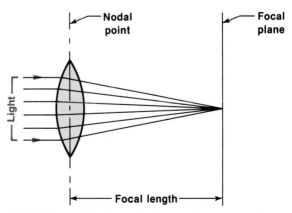

FIGURE 13-5. Focal length of a lens is measured from the center (nodal point) of the lens to the focal plane.

It is more difficult to determine the center of a multiple element (compound) lens because this point can vary. The number of elements, their shapes, and grouping are factors affecting the nodal point location. The scientific definition of focal length includes the statement: "FL = the distance from the *rear nodal point* of the lens to the focal plane."

Focal length is always given in millimeters. The standard lens supplied with most 35-mm SLR cameras has a focal length of 50 mm. This can vary slightly, depending on the camera brand, from 47 to 55 mm. This lens gives about the same relative image size as one human eye.

ANGLE OF VIEW

The focal length of a lens determines its *angle of view*, Figure 13-6. The shorter the focal length of the lens, the wider its angle of coverage. The longer the focal length of the lens, the narrower the angle of coverage.

The perspective and size of the image refracted through the lens also are determined by its focal length. Examples show how the image content changes by using lenses of different focal lengths, Figure 13-7. Lenses for 35-mm cameras with focal lengths less than 40 mm are called wide angle and those above 60 mm are called telephoto.

Zoom lenses make it possible to obtain different angles of view and perspective with one lens, Figure 13-8. Selected elements within the lens barrel move to give the various focal lengths within the range of the lens.

Special lenses, such as the *fisheye* lens, have an angle of view of 180 degrees. Images are not accurately represented, but this type of lens is useful for special effects. Macro lenses increase the

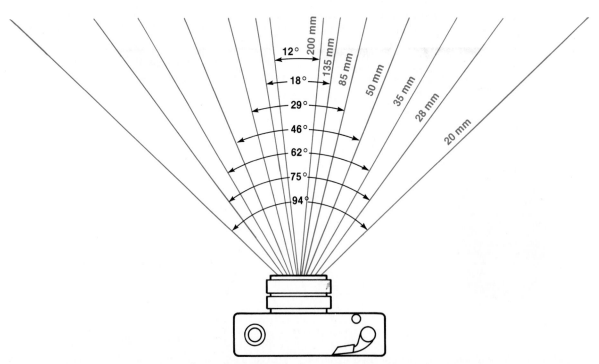

FIGURE 13-6. The angle of view of a lens is determined by its focal length.

30mm

48mm

80mm

250mm

600mm

800mm

FIGURE 13-7. A comparison of lenses of six focal lengths. *Tamron Industries, Inc.*

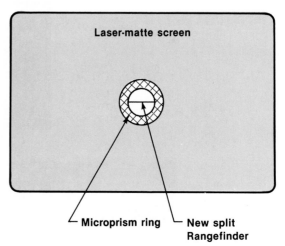

Laser-matte screen

Microprism ring   New split Rangefinder

FIGURE 13-9. The split-image rangefinder is useful for rapid focusing.

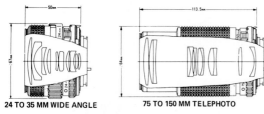

24 TO 35 MM WIDE ANGLE   **75 TO 150 MM TELEPHOTO**

FIGURE 13-8. Zoom lenses provide the photographer with a selection of focal lengths. *Minolta Corporation*

image size while maintaining a given focal length and angle of view. These lenses are excellent for close-up photography and for portraits.

FOCUSING A LENS

Several types of *focusing screens* (see Figure 9-7) are available for SLR cameras. Several screens contain a center area called the *split-image* rangefinder, Figure 13-9. To focus, select a straight line area on the subject. Move the lens forward and back until the image line is joined both above and below the split-image line in the viewfinder. The micro-

prism ring and the matte screen also can be used to focus the lens. Low light conditions sometimes make these two areas difficult to use.

Automatic focusing is available on some cameras. One system uses a sound system. In this system, a "chirp" of ultrasound, which cannot be heard by humans, is emitted from the camera. A special quartz clock and a computer within the camera measure the amount of time it takes for the echo of the ultrasonic chirp to return from the subject. Once done, a tiny electric motor moves the lens to the focused position. Another system uses infrared light, while another system uses two matching mirrors and one prism. When reflected light from the subject is equal in both mirrors, the camera computer tells the lens motor where to position the lens.

Computer-aided systems also help the photographer focus the lens. LEDs are positioned on the edge of the viewing screen where they can be seen easily, Figure 13-10. For example, if the left LED is lighted, the lens should be turned to the right—greater distance. When the center LED lights, the image is in focus and the picture can be taken.

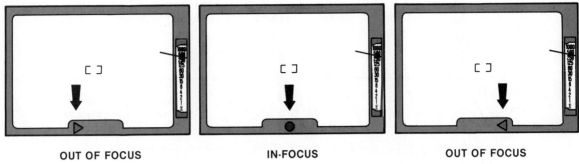

OUT OF FOCUS   **IN-FOCUS**   **OUT OF FOCUS**

FIGURE 13-10.   LEDs tell the photographer whether the lens is focused in the computer-aided focusing system.

DEPTH OF FIELD

Depth of field is defined as the distance between the nearest point and the farthest point that is in acceptable focus when a lens is directed at three-dimensional subjects, Figure 13-11. There are three methods of controlling and changing depth of field:

1. Changing the aperture opening (f-stop)
2. Changing the subject-to-camera distance
3. Changing the lens focal length, Figure 13-12

The most common control is that of the aperture opening, which is more commonly called the f-stop (see unit 15). A smaller aperture will always give a greater depth of field. Markings on lens barrels provide helpful information as to the depth of field of a given f-stop, Figure 13-13. The depth-of-field scale contains f-stop markings on both sides of the center line. When a lens aperture is adjusted for a specific f-stop, it is convenient to determine the depth of field. In the illustration, the f/8 setting places all image content within the 8 to 14 feet (2.4 to 4.2 m) that are in focus. The f-stops of f/11 and f/16 give even greater depth-of-field ranges.

It is often useful to select an object that is one third into the depth-of-field range of the selected f-stop. Approximately two thirds of the total depth-of-field range will be behind this spot. This occurs due to the optics of the lens. The *hyperfocal* distance is a useful tool for the photographer.

It is defined as the minimum distance that is in focus when a lens is adjusted to infinity at a specified f-stop on the depth-of-field scale.

LENS ABERRATIONS

All lenses contain defects that affect the ability of the lens to correctly refract light. These defects are

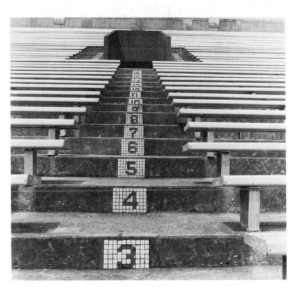

FIGURE 13-11.   A small f-stop and slow shutter speed were used to obtain the depth of field in this stadium photograph. *Barton A. Dennis*

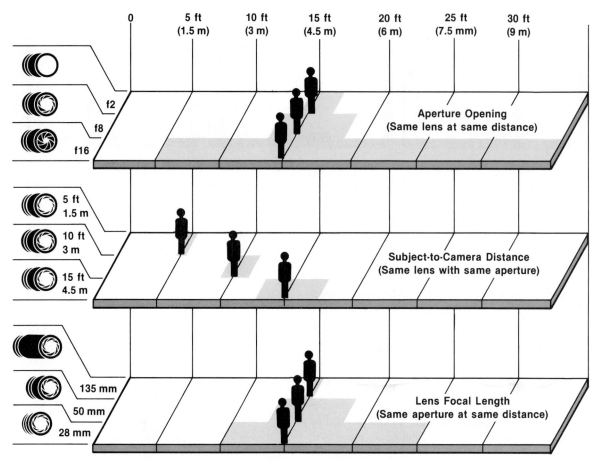

FIGURE 13-12. The three methods of controlling and using depth of field.

called *aberrations*. In most cases, they are so minor that the photographer need not worry about them. Lens manufacturers continue to find ways of reducing these defects.

LENS ACCESSORIES

Several devices and supplementary lenses have been developed to increase the image size on film. Among these accessories are extension tubes, bellows, supplementary lenses, and teleconverters.

Extension tubes and *bellows* are designed to be positioned between the camera and the lens. Extension tubes are available in different sizes or thicknesses. A typical set of three tubes includes 12, 25, and 36 mm sizes, Figure 13-14. One or more used in combination are locked in place on the bayonet lens mount of the camera. The lens, usually the standard 50 mm, then is attached on the bayonet front end of the extension tube. These tubes place the lens farther from the film; thus, the image refracted through the lens is enlarged when it reaches the film.

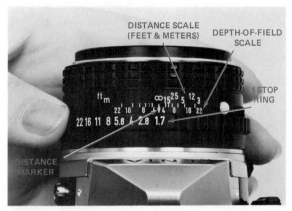

FIGURE 13-13. The depth-of-field scale stamped in the lens barrel is useful when selecting the f-stop. *Pentax Corporation*

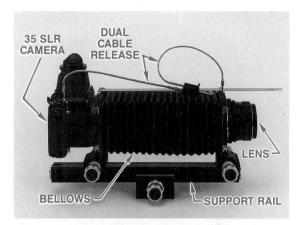

FIGURE 13-15. A bellows unit between the camera and lens allows continuous distance adjustment. *Pentax Corporation*

FIGURE 13-14. A standard set of extension tubes. *Pentax Corporation*

FIGURE 13-16. Supplementary lenses fasten directly to the front of a regular camera lens. *Pentax Corporation*

Bellows do the same things as extension tubes, Figure 13-15. The advantage of bellows over extension tubes is that there is a continuous adjustment between the minimum and maximum lens positions.

Supplementary lenses and *teleconverters* are also designed to increase the image size on the film while continuing to use a regular camera lens. These devices, though, decrease the speed of the lens. This allows less light to pass through the lens. High-speed film can be used to offset this slight

problem. These lenses are identified as $+1$, $+2$, $+10$, $+20$, depending upon the amount of magnification each provides, Figure 13-16. Supplemen-

FIGURE 13-17. Teleconverters give extra image magnification with little change in the camera-to-subject distance. *GMI Photographic*

FIGURE 13-18. The manufacture of photographic lenses requires considerable hand labor. *Tamron Industries, Inc.*

tary lenses provide an inexpensive way to obtain slightly larger images.

Teleconverters increase image magnification without appreciably changing the distance from the front of the lens to the subject, Figure 13-17. This simplifies the placement of filters and flash units compared to using tubes or bellows extensions to gain similar image magnifications. The magnification increase is arithmetically proportional to the power of the converter. Thus, a macro lens capable of making a 1:1 (life-size) image unaided will render an image 2:1 (twice life-size) with a 2X converter and 3:1 with a 3X one.

LENS MANUFACTURE

Lenses of all types and sizes are precision devices. They must be manufactured and assembled under exacting conditions. It takes considerable hand labor to prepare lenses for use in cameras and other photographic equipment, Figure 13-18. The working environment must be kept very clean so dust and dirt particles do not damage the lens surfaces.

SUMMARY

Lenses are the eyes of a camera. They are designed to gather, focus, and transmit light to the camera focal plane. At this point, the film emulsion captures the image being refracted through the lens. Because there are many lens types and configurations, it is possible to capture a wide variety of scenes on film. Camera lenses are very critical to photography. Select and use a lens with utmost respect and care.

REVIEW QUESTIONS

Answer these questions to test your knowledge of the unit content.

1. Why is the lens a critical component of a camera?
2. Light is bent or _____ as it passes through a lens.
3. Which of the following selections is *not* a basic surface shape of a lens?

   A. Convex   C. Flat
   B. Irregular   D. Concave

4. T/F The diverging or minus type of lenses cause light to be directed to a small area.

5. T/F Lenses for 35 mm SLR cameras can contain up to 16 individual elements.

6. Focal length of a lens is measured from the _____ to the focal plane.

7. The angle of view of which lens is about that of the human eye?

A. 200 mm   C. 80 mm
B. 35 mm   D. 50 mm

8. Name three focusing methods currently being used on adjustable cameras.

9. Which method of controlling depth of field can be done with a simple on-camera adjustment?

A. Aperture change
B. Film speed change
C. Lens focal length
D. Subject-to-camera distance

10. Which lens accessory provides the greatest amount of adjustment when photographing very small images?

A. Extension tubes   C. Bellows
B. Teleconverter   D. Auxiliary lens

--- UNIT 14 ---

CONTROLLING EXPOSURE

OBJECTIVES   *Upon completion of this unit, you will be able to:*
- *Identify the standard f-stops found on the lens of an adjustable camera.*
- *Compute the f-stop rating of a lens given the lens length and the maximum aperture diameter.*
- *Name and describe the two types of shutters used on cameras.*
- *Select the best aperture and shutter speed combination for given situations.*

KEY TERMS   *The following new terms will be defined in this unit:*

Aperture	f-stop	Leaf Shutter
Aperture Ring	Half-stop	Shutter
Focal-plane Shutter	Iris Diaphragm	

INTRODUCTION

Many variables must be controlled before quality photographs can be made. This is true at each stage of the photographic process. The major variable that the photographer must control is the amount of light that reaches the film. Adjustable cameras contain two major controls, lens aperture and shutter speed, which photographers must know how to handle.

LENS APERTURES

An *aperture* is an opening which lets something pass through. A lens aperture allows light to pass through a lens carrying the image to the film within the camera. Numbers called *f-stops* are used to measure the size of the aperture openings, Figure 14-1. The larger the opening, the more light it will pass.

An *iris diaphragm* is used to regulate the aperture openings. It is a series of metal leaves that

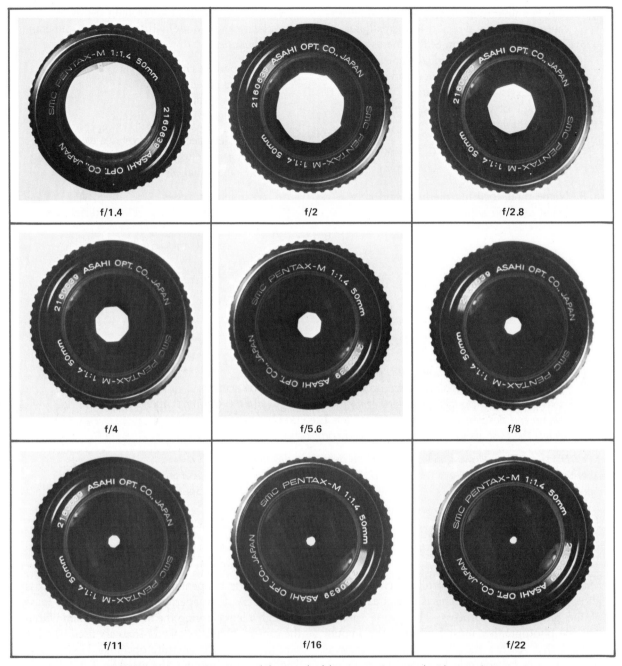

FIGURE 14-1. A comparison of the standard f-stop openings used with most lenses.

are accurately controlled by an intricate system of pins and linkages, Figure 14-2. The iris diaphragm is located within a lens between the elements of ground and polished glass. Six to eight metal leaves are used to make a typical iris diaphragm.

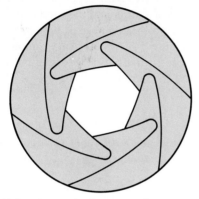

FIGURE 14-2. The iris diaphragm used in a camera lens is a series of precise overlapping metal leaves.

The numbers used to designate given apertures of f-stops are based on the diameter of the diaphragm opening and the focal length of a lens. For example:

Lens focal length = 50 mm
Maximum diaphragm opening = 35 mm

The f-stop number is obtained by dividing the focal length by the maximum diaphragm opening. Only one decimal place is used, otherwise it would become extremely confusing.

$$\frac{50 \text{ mm}}{35 \text{ mm}} = 1.429 \text{ thus } f/1.4$$

As the diaphragm opening becomes smaller, the f/stop number becomes larger. For example:

$$\frac{50 \text{ mm}}{6.25 \text{ mm}} = 8.0 \text{ thus } f/8$$

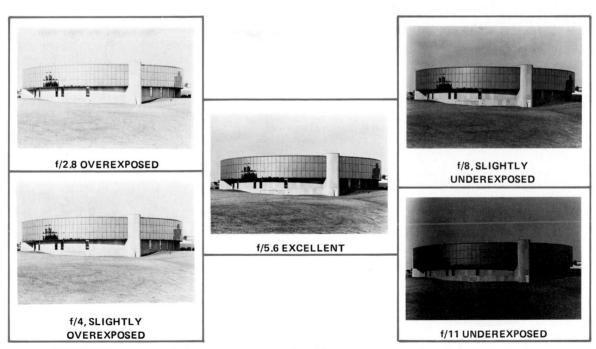

f/2.8 OVEREXPOSED

f/4, SLIGHTLY OVEREXPOSED

f/5.6 EXCELLENT

f/8, SLIGHTLY UNDEREXPOSED

f/11 UNDEREXPOSED

FIGURE 14-3. Prints made from negatives exposed at different apertures on the standard f-stop system.

The f-stop system has been designed so either twice as much or half as much light passes through the aperture at the adjacent f-stop. For instance, f/2.8 lets in twice as much light as f/4. The reverse is true in that f/4 lets in half as much light as f/2.8. This relationship of adjacent f-stops holds true anywhere on the standard f-stop scale, (see Figure 14-1). The selection of f-stops does make a difference in the amount of light each opening allows through a lens, Figure 14-3.

Half-stops are the aperture adjustments midway between the standard f-stops. These are useful when just a little more or less light is needed to make the correct exposure.

The lens *aperture ring* is used to adjust the diaphragm for the various f-stops, Figure 14-4. The f-stops are clearly marked, generally in white, on the ring. This makes them easy to see against the black color of the lens barrel. A bold setting line, usually white, is marked in the stationary area of the lens barrel next to the aperture ring. An f-stop is set by aligning the selected f-stop number and the bold white line.

Exact f-stops, whether full or half, are easy to adjust. The aperture ring contains "clicks" or slight notches that can be felt through the fingers when the ring is moved. The clicks also can be heard with most lenses.

CAMERA SHUTTERS

A camera *shutter* is similar to a door. It is closed, part open, fully open, part closed, and again closed. During this cycle, a specific amount of light is permitted to reach the film.

There are two basic types of shutters used in cameras: the leaf shutter and the focal-plane shutter. Leaf shutters are located near or within the lens of a camera, Figure 14-5. Focal-plane shutters are located as close to the film as possible but between the lens and the film, Figure 14-6.

THE LEAF SHUTTER

A *leaf shutter* is made of three or more very thin blades. The material used to make the blades is

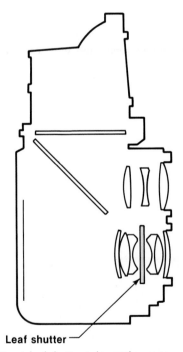

Leaf shutter

FIGURE 14-5. A leaf shutter is located very near or within the camera lens.

— Setting Line

— Aperture Ring

FIGURE 14-4. The f-stops and setting line are clearly marked on the lens barrel. *Tamron Industries, Inc.*

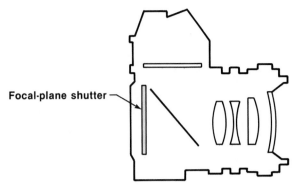

FIGURE 14-6. A focal-plane shutter is located just ahead of the film.

Focal-plane shutters are designed to operate either in a vertical or horizontal direction across the film frame.

Two curtains move or run across the film frame, one ahead of the other. When the shutter release is squeezed, the lead curtain travels in front of the film and is then followed shortly thereafter by the trailing curtain. The shutter speed setting determines how fast the curtains move and how much space or opening is between the curtains. With a slow shutter speed of 1/125 or less, the shutter uncovers the film completely. When the shutter speed is 1/250 or faster, the film is exposed in a narrow band, Figure 14-9. This is because the trailing curtain follows the lead curtain so rapidly.

Shutter speeds are measured in fractions of a second. A 1-second shutter speed is very slow, whereas a 1/2000th of a second shutter speed is very fast. Shutter speeds are marked on the camera only with the denominator of the fraction. For example, a marking of 250 equals 1/250 of a second.

either spring steel, plastic, or titanium. A series of "clock-like" precision parts make it possible for the blades to open and close, Figure 14-7. Leaf shutters are used on nearly all viewfinder cameras because they cost less than focal-plane shutters. Some medium format cameras have leaf shutters built into the interchangeable lenses. Leaf shutters also are used in twin-lens reflex and view cameras.

THE FOCAL-PLANE SHUTTER

Focal-plane shutters are used in nearly all 35-mm, single-lens reflex cameras, Figure 14-8. These shutters include two rubberized fabric or metal curtains that are mounted as close to the film as possible.

SHUTTER SPEED CHOICES

Many 35-mm SLR cameras are now equipped with up to 14 choices of shutter speeds, Figure 14-10. This gives the photographer a wide variety of choices. Shutter speed adjustments and readings are commonly located in two different areas of 35-mm SLR cameras. Manual adjustable cameras have a shutter speed dials often located on the top and

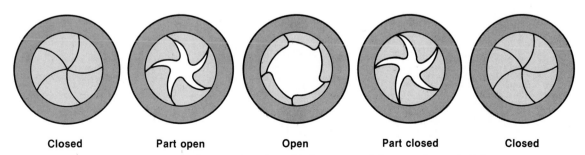

| Closed | Part open | Open | Part closed | Closed |

FIGURE 14-7. Precision thin metal or plastic blades open and close rapidly in a leaf shutter.

right side near the film advance. Numerical values, LCDs, and LEDs are used in the viewing screens of automatic and program cameras to indicate the shutter speed selection, Figure 14-11.

Adjacent shutter speeds, like f-stops, halve or double the amount of light that passes through the lens. For example, a 1/125 shutter speed permits half the light as does 1/60. The same 1/125

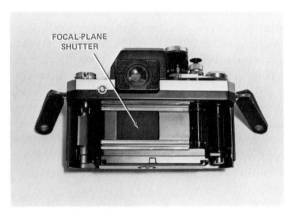

FOCAL-PLANE
SHUTTER

FIGURE 14-8. Focal-plane shutters are made of two curtains.

shutter speed setting permits twice as much light to pass through the lens as does 1/250. This principle is basically true at all camera shutter speeds. There are some slight differences between the halving and doubling of transmitted light with some shutter speeds. This does not often create a problem unless critical work is being photographed.

Shutter speed and f-stop adjustments have a direct relationship. The several shutter speed and f-stop combinations permit the same amount of light to reach the film during exposure, Figure 14-12. For example, a 1/125 shutter speed and an f/11 combination permits the same amount of light to pass through as 1/250 shutter speed and f/8.

SUMMARY

Aperture openings regulate how much light passes through a lens at any given time. The various aperture openings are identified by f-stop numbers that typically range from f/1.4 (largest) to f/22 (smallest). The second major exposure control is shutter speed. The two types of shutters are leaf and focal-plane. The 35-mm SLR camera is typically designed with a focal-plane shutter. Adjacent f-stops and

Slow shutter speed

Fast shutter speed

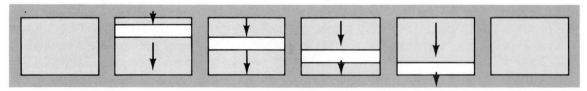

FIGURE 14-9. Two curtains move across the film frame to expose the film with a focal-plane shutter.

A Typical Shutter Speed Series	
4s = 4 seconds	30 = 1/30 second
2s = 2 seconds	60 = 1/60 second
1 = 1 second	125 = 1/125 second
2 = ½ second	250 = 1/250 second
4 = ¼ second	500 = 1/500 second
8 = 1/8 second	1000 = 1/1000 second
15 = 1/15 second	2000 = 1/2000 second

FIGURE 14-10. The photographer has the choice of many shutter speeds with 35-mm SLR cameras.

shutter speeds permit half or twice the amount of light to pass through a lens. Several combinations of f-stops and shutter speeds transmit the same amount of light.

REVIEW QUESTIONS

Answer these questions to test your knowledge of the unit content.

1. Why must photographers learn how to control light?

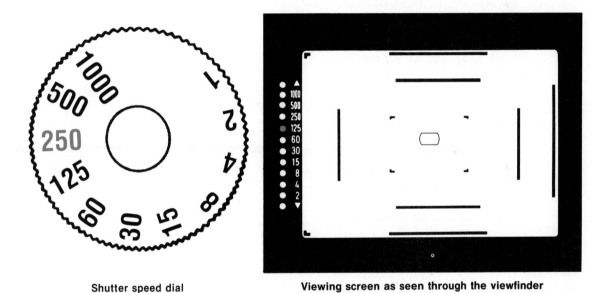

Shutter speed dial   **Viewing screen as seen through the viewfinder**

FIGURE 14-11. The two common locations where shutter speed adjustments are made and observed.

f-stop	22	16	11	8	5.6	4	2.8	2	1.4
Shutter Speed	1/8	1/15	1/30	1/60	1/125	1/250	1/500	1/1000	1/2000

FIGURE 14-12. Each of these f-stop and shutter speed combinations permits the same amount of light to reach the film.

2. T/F An f-stop is a specific location on the aperture ring of a lens.

3. Where is the iris diaphragm located on an adjustable camera?

   A. Behind the lens elements
   B. Ahead of the lens elements
   C. Between the lens elements
   D. Just ahead of the focal plane

4. A 50-mm lens with a diaphragm opening of 20 mm would have an f-stop of:

   A. f/11   C. f/4
   B. f/32   D. f/2.5

5. A 50-mm lens set at f/11 would have a diaphragm opening of:

   A. 4.5 mm   C. 2.5 mm
   B. 6.0 mm   D. 7.2 mm

6. T/F The next larger full f-stop on a standard lens permits 50% more light to pass through.

7. T/F The photographer can easily determine when a specific f-stop adjustment has been obtained.

8. A leaf shutter is made of very thin metal or plastic _____.

9. Focal-plane shutters are most often used in:

   A. TLR cameras
   B. View cameras
   C. SLR cameras
   D. Viewfinder cameras

10. T/F Programmed cameras give a limited number of shutter speed selections.

──────── UNIT 15 ────────

CHOOSING FILM

OBJECTIVES   *Upon completion of this unit, you will be able to:*
   • *Select suitable film for the pictures to be taken.*
   • *Explain the different film speed rating systems.*
   • *Predict the increase or decrease of visible grain.*
   • *Apply knowledgeable film buying practices.*

KEY TERMS   *The following new terms will be defined in this unit:*

ASA	EI	ISO
Bulk Loader	Film Contrast	Light-tight
Bulk Loading	Film Grain	Pushing (film)
DIN	Film Speed	Visible Grain

INTRODUCTION

Photographic film is the material used to capture the scene. Several brands, types, and speeds are available from retail and catalog stores. Choosing the right film for the situation is important because the choice of film has a significant effect on the outcome of the photograph.

FILM SPEED

It is important to select a film that is suitable for the pictures to be taken. Probably the most important consideration is the speed of the film. All but a few continuous-tone photographic films are rated according to their sensitivity to light. The system is known by the initials *ISO* (International Stan-

ISO	25	32	40	50	64	80	100	125	160	200	250	320	400	500	650	800	900	1000
DIN	15	16	17	18	19	20	21	22	23	24	25	26	27	28	29	30	31	32

FIGURE 15-1. A comparison of the ISO and DIN film speed numbering systems.

dards Organization). Typical film speed ratings are ISO 64, ISO 125, ISO 200, ISO 400, and ISO 1000.

Another film speed identification system is known by the three initials *DIN* (Deutsche Industrie Norm). Translated from German to English, it means German Industrial Standard. Typical DIN ratings include 15 through 32. See the comparison chart, Figure 15-1.

It is very important to adjust the camera film speed dial to the ISO rating of the film, Figure 15-2. With this information, the light meter within the camera will be able to calculate the correct f-stop and shutter speed settings.

The ISO system was formerly known by the initials *ASA* (American National Standards Institute). Now, though, only ISO numbers are used for films sold in America; whereas, DIN film numbers are used in most countries of Europe. The common practice is to list both speed rating systems, for example, ISO 400/27°.

Film speeds are easy to understand. Assuming that ISO 25 is the base, a film with an ISO number of 50 is twice as sensitive to light. Film having an ISO number of 125 is five times as sensitive. Camera shutter speeds and f-stops also enter into the film speed rating system. Increasing the aperture opening one full stop doubles the amount of light reaching the film. The same is true when the shutter speed is decreased one setting (see unit 14). Thus, using a film with an ISO rating twice as fast as a previous film allows the next aperture or f-stop to be used, Figure 15-3.

Generally, speed ratings lower than ISO 100/21° are considered to be slow films. Films of ISO 400/27° and higher are listed as fast films. Medium speed films fall between these numerical values.

FILM GRAIN

Every piece of processed photographic film has visible grain. The object is to obtain acceptable grain in the film negative and resultant prints. *Visible grain* is when the developed silver halide crystals

FIGURE 15-2. Adjusting the camera film speed dial to the ISO rating of the film.

Shutter	1/125	1/125	1/125	1/125	1/125	1/125
f-stop	4	5.6	8	11	16	22
ISO	32	64	125	200	400	1000
Shutter	1/60	1/125	1/250	1/500	1/1000	1/2000
f-stop	5.6	5.6	5.6	5.6	5.6	5.6

FIGURE 15-3. Each of these ISO numbers (film speed ratings), shutter speeds, and f-stops give the same picture-taking results.

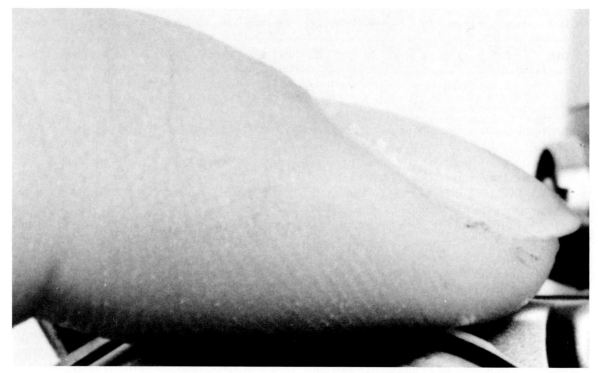

FIGURE 15-4.   A greatly enlarged portion of a photograph showing visible grain. *Pentax Corporation*

making up the film emulsion can be seen in the photographic print, Figure 15-4.

The standard cause of visible grain is the ISO speed of the film. The higher the speed, the larger the silver halide crystals making up the light-sensitive portion of the film (see unit 24). With this in mind, it is advisable to use the lowest speed film as possible based upon the available lighting conditions. Manufacturers have made considerable progress in recent years in reducing visible grain in highspeed films. Photographic examples of slow, medium, and fast films are shown, Figure 15-5.

FILM CONTRAST

Contrast is the difference between the lightest tonal areas (highlights) and darkest tonal areas (shadow) of a negative or a photograph. Contrast within the image actually makes the photographic content visible. Photographs with higher contrast are usually more desirable, especially for reproduction by one of the graphic arts printing processes in books, magazines, newspapers, and other printed products. Slow films generally provide for less contrast between the highlight and shadow areas of the image. Fast films tend to have higher contrast, Figure 15-6.

PUSHING FILM

Sometimes there is need to have a higher speed film than what is available. For example, a film with an ISO rating of 200 can be doubled to an ISO rating of 400. This is simply done by adjusting the film speed setting on the camera. After the entire roll is exposed, it must be developed according to instructions provided by the company manufac-

| SLOW, ISO 32/16° | MEDIUM, ISO 125/22° | FAST, ISO 400/27° |

FIGURE 15-5. Portions of 8 × 10 photographs made from 35-mm negatives. Notice the visible grain as the speed of the film was increased.

turing the film. See unit 29 for the special processing needed when pushing film. It is possible to push the film even more than twice its standard speed. Whenever film is "pushed," some quality is lost. It should only be done whenever it is absolutely necessary. Also, only selected films can be pushed. See the instructions with the film.

The term "Exposure Index" (*EI*) is used to designate that a film has been pushed. For example, if a film with an ISO of 400 is exposed at 800, it should be referred to as EI800 rather than ISO 800. This makes it clear that the film was exposed for more than its rated speed.

FILM BRANDS AND TYPES

Several companies manufacture photographic film throughout the world. It is important to select a film brand and type that fits the needs of the photographer and the situation. If there is a sufficient amount of light, it is generally wise to select a slow to medium speed film. Low light and/or moving objects may require a fast film. The important point is to obtain the right film for each and every occasion, Figure 15-7. There are a wide variety of films; thus, read the information on the packaging carefully.

When loading a camera, look closely at the film cartridge or magazine, Figure 15-8. Important information is printed on the *light-tight* container. The manufacturer has made every effort to eliminate confusion.

COLOR OR BLACK-AND-WHITE

The world is full of color, and many people believe all photographs should be made with color film. Color has its important advantages as shown in the several units of Chapter 7. Color films for either slides or prints are available from most retail outlets. Color film designed for photographic prints most

SLOW, ISO 32/16°   FAST, ISO 400/27°

FIGURE 15-6. Contrast between the highlights and the shadows is usually greater with a high speed film.

often contains the word "color" in its title. Film designed for slides most often contains the word "chrome" in its title. Also, additional information can be found on the film package and light-tight container that states either, "for color slides," or "for color prints."

Some color films give the photographer a choice of either being processed for negatives (prints) or for positives (slides). By exposing the processed negatives to an unexposed roll of film, it is possible to obtain a set of slides.

Black-and-white photography is useful in many ways. This is especially true for photographs

that will be used in printed products. Thus, the photographer must select the correct film for the intended use.

BULK LOADING

Most brands, types, and speeds of film can be purchased in bulk amounts of 100 feet (30.5 m) rolls. *Bulk loaders* make it fast and easy to fill reusable cassettes with film for use in the camera, Figure 15-9. Reusable film cassettes for 35-mm film are designed to hold enough film for up to 36 exposures. Using bulk loading saves money on

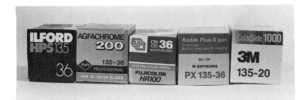

FIGURE 15-7. Valuable information is printed on the film boxes, such as the brand, type, and speed of the film.

FIGURE 15-8. Critical information also is printed on the film containers so there will be no confusion.

FIGURE 15-9. A bulk film loader used to load a reusable 35-mm film magazine. *Queen City Plastics*

the purchase of film and permits the photographer to load different amounts of film in the cassette. This too saves on film cost. When bulk loading film, it is important not to put more than 36 exposures in a roll. There is not sufficient room for any more. Also, fasten the end securely and mark cassette cartridges plainly with brand, film type, speed, and number of exposures.

SUMMARY

Choosing the right film for the occasion is important. There are many film speeds available in both black-and-white and color film. Different companies manufacture film, giving the photographer a choice of several brands and types. When large amounts of film will be used, bulk purchase and loading can be very economical

REVIEW QUESTIONS

Answer these questions to test your knowledge of the unit content.

1. Which film characteristic is probably the most important to consider?

   A. Grain   C. Exposure index
   B. Speed rating   D. Contrast

2. The initials ISO are the abbreviation for _____.

3. Identify the film speed rating system that is based on German standards.

   A. DIN   C. ISO
   B. NID   D. ASA

4. Select the most common film speed identification printed on film packaging.

   A. DIN 200/24°   C. ISO 200/24°
   B. NID 200/24°   D. ASA 200/24°

5. If there is need to use a faster shutter speed than a given film will permit, the photographer can:

   A. Try a different camera
   B. Use a faster film
   C. Obtain a new lens
   D. Set the camera on automatic

6. Visible grain in a photograph is generally caused from the:

   A. High ISO of film
   B. Photographic paper
   C. Chemical processing
   D. Shape of the lens

7. T/F Higher contrast between tonal areas in a photograph is most often achieved by using a slow speed film.

8. Exposure index and _____ have a direct relationship.

   A. Developing film
   B. Pushing film
   C. Amount of visible grain
   D. Film contrast

9. Which information must be printed on film packaging so the photographer can set the camera correctly?

   A. Color or black-and-white
   B. Brand
   C. Size
   D. Speed

10. T/F Color film manufactured for photographic prints frequently contains the word "chrome" in its title.

UNIT 16

MEASURING LIGHT

OBJECTIVES   *Upon completion of this unit, you will be able to:*
  • *Describe through-the-lens (TTL) metering.*
  • *Adjust a manual camera for a proper exposure.*
  • *Use cameras with computer-aided light metering.*
  • *Describe and use light meters to determine correct f-stops and shutter speeds.*

KEY TERMS   *The following new terms will be defined in this unit:*

Aperture Priority	*Incident Light*	*Selenium*
CdS	*Light Meter*	*Shutter Preselection*
Computer-Aided Metering	*Manual Override*	*Shutter Priority*
EVS	*Photocell*	*SPD*
Exposure Meter Index	*Programmed Camera*	*Spot Meter*
Exposure Meter Needle	*Reflected Light*	*TTL*
f-stop Preselection		

INTRODUCTION

Measuring the amount of light falling on a scene is critical to quality photography. Once the quantity of light has been determined, it is possible to accurately set the f-stop opening and the shutter speed. Precise instruments called *light meters*, Figure 16-1, are used to measure the light and provide important data.

TTL METERING

The initials *TTL* stand for "through-the-lens." This type of metering is built into single-lens reflex (*SLR*) cameras and in some viewfinder cameras. It provides for convenience and accuracy in adjusting the f-stop and shutter to the correct settings.

TTL metering permits light to be measured at several points after it has passed through the lens. One of the common locations on manual adjusting SLR cameras is on the inside of the eyepiece. Highly sensitive *photocells* measure the light and transmit it to a meter needle visible through the viewfinder, Figure 16-2. The *exposure meter index* informs the photographer whether the exposure settings are correct, over, or under.

MANUAL ADJUSTMENTS

The following procedure can be used to manually set an SLR camera to accurately record the scene being photographed.

1. Select the scene.
2. Preselect the f-stop or shutter speed.
  * *f-stop preselection*—Set the f-stop according to the scene and lighting conditions. If greater depth of field is desired, a small aperture opening must be used. If the lighting is low, a larger aperture opening will need to be used.
  * *Shutter preselection*—Set the shutter speed according to the conditions of the scene. It will be necessary to use a fast shutter of 1/500 or above for bright light conditions. If the lighting is low, it would be well to select a shutter of 1/60 or 1/125.

3. Activate the metering system. Some cameras contain off–on switches. On others, a slight depression of the shutter release button will activate the metering system.

4. Look through the viewfinder and observe the *exposure meter needle*. On most cameras, the objective is to place the needle in the center of the exposure meter index, Figure 16-3.
  * For f-stop preselection, rotate the shutter speed ring until the needle lines up as close as possible to the center of the index.
  * For shutter preselection, turn the aperture ring on the lens barrel back and

FIGURE 16-1.   A hand-held light meter being used to accurately measure the light reflecting from a scene.

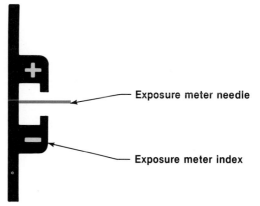

Exposure meter needle

Exposure meter index

FIGURE 16-2.   On manually adjustable SLR cameras, the meter needle is visible in the viewfinder. *Olympus Corporation*

forth until the needle is centered in the index.

5. Squeeze the shutter release button and take the picture.

INTENTIONAL OVER- OR UNDEREXPOSURES

Over- or underexposures can be made to meet special lighting requirements. Sometimes the back lighting or side lighting of the scene requires that aperture or shutter settings be different from the normal "correct" exposure adjustments. The exposure meter index and needle can be used as a guide, Figure 16-4. When the needle swings toward the (+) position, it indicates overexposure or too much light will reach the film. When the needle swings towards (−), it indicates underexposure.

COMPUTER-AIDED METERING

Many cameras with TTL metering contain microcomputers that measure the light and automatically

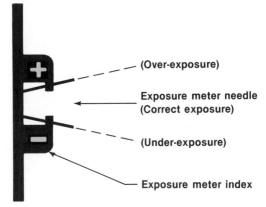

FIGURE 16-3.   For a correct exposure, the exposure meter needle must be in the center of the exposure meter index. *Olympus Corporation*

make camera settings, Figure 16-5. This includes setting the aperture opening or the shutter speed, or both the aperture and shutter. The central processing unit (CPU) functions as the brain of a camera's sophisticated electronic system. Cameras that select either the aperture or shutter are considered to be automatic. Those that select both are called *programmed.*

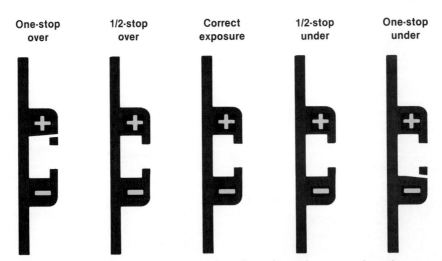

FIGURE 16-4.   The exposure meter index and needle can be used to accurately predict over- and under-exposure. *Olympus Corporation*

FIGURE 16-6. The LED in the viewfinder of an aperture priority camera signals the shutter speed that was selected by the microcomputer.

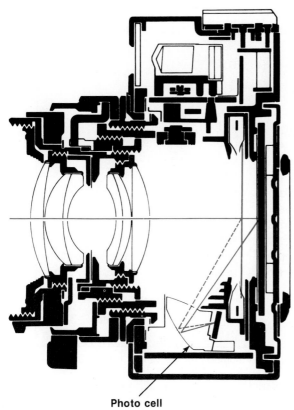

Photo cell

FIGURE 16-5. The highly sensitive silicon photocell measures the light that strikes the shutter curtains or the film itself.

APERTURE AND SHUTTER PRIORITY

Aperture priority cameras require that the photographer first sets the aperture (f-stop) according to the scene conditions. From this point, the microcomputer uses the information from the TTL light meter and selects the correct shutter speed, Figure 16-6. The picture can be taken immediately after the lens is focused and the aperture is selected. This saves the photographer considerable time and effort.

Shutter priority cameras require just the reverse of the aperture priority cameras. The photographer needs only select the shutter speed, focus the lens, and squeeze the shutter. The correct f-stop

will be selected by the central processing unit within the camera body.

The photographer needs to decide which camera priority is best for the type of pictures that most often will be taken. If depth of field is an important consideration with a majority of the photographs, then an aperture priority camera should be purchased. If stop-action scenes are common practice, then it would be well to have a shutter priority camera.

PROGRAMMED CAMERAS

Cameras with this advanced feature make it easy to take pictures. The photographer needs only to select the scene, focus the lens, and squeeze the shutter button.

A typical programmed camera is based upon the exposure value system (*EVS*). This is sometimes abbreviated to EV. This system gives a standard number to a selected combination of f-stop and shutter speed, Figure 16-7. An EV 13 indicates an exposure of f/2 at 1/1000; whereas, an EV 19 calls for f/16 at 1/1000. Other combinations also are available for the same EV number depending on the f-stop or shutter speed desired. The EVS is not widely used.

EV No.	Apertures: f/2	f/2.8	f/4	f/5.6	f/8	f/11	f/16	f/22	f/32
2	1								
3	1/2	1							
4	1/4	1/2	1						
5	1/8	1/4	1/2	1					
7	1/15	1/8	1/4	1/2	1				
8	1/30	1/15	1/8	1/4	1/2	1			
9	1/60	1/30	1/15	1/8	1/4	1/2	1		
10	1/125	1/60	1/30	1/15	1/8	1/4	1/2	1	
11	1/250	1/125	1/60	1/30	1/15	1/8	1/4	1/2	1
12	1/500	1/250	1/125	1/60	1/30	1/15	1/8	1/4	1/2
13	1/1000	1/500	1/250	1/125	1/60	1/30	1/15	1/8	1/4
14		1/1000	1/500	1/250	1/125	1/60	1/30	1/15	1/8
15			1/1000	1/500	1/250	1/125	1/60	1/30	1/15
16				1/1000	1/500	1/250	1/125	1/60	1/30
17					1/1000	1/500	1/250	1/125	1/60
18						1/1000	1/500	1/250	1/250
19							1/1000	1/500	1/125
20								1/1000	1/500
21									1/1000

FIGURE 16-7. The EVS (exposure value system) table gives the EV numbers for the various f-stop and shutter speed combinations.

The microcomputer within the camera selects the f-stop and shutter speed based on information preprogrammed during manufacture, Figure 16-8. The camera is generally programmed to use the fastest shutter speed that the lighting will allow. The programmed information even includes adjustments for different lens f-stop ratings.

MANUAL OVERRIDE

Most of the automatic and programmed cameras have this valuable feature. The manual mode permits the photographer to select both the f-stop and shutter speed. This feature is especially useful when the subject of the scene is backlit.

When there is considerable light in the background, the TTL light meter will be fooled. This will cause the subject to be too dark, Figure 16-9. To correct this problem, it is necessary to take a close-up light meter reading, Figure 16-10. Then set the camera to manual mode, adjust the f-stop or shutter speed, and step back and take the picture. The subject will be properly exposed, and the background will be slightly overexposed, Figure 16-11. Some cameras have a "memory lock" feature that remembers the close-up meter reading until the shutter is released.

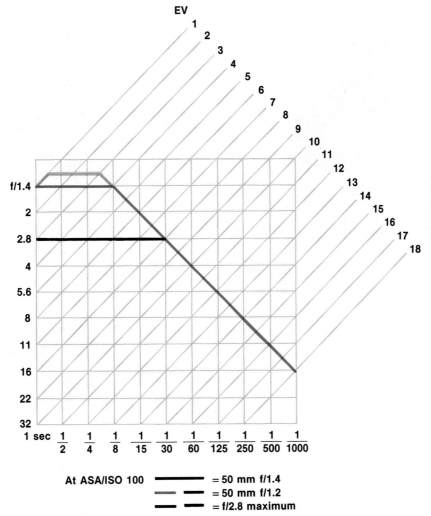

FIGURE 16-8. A graph showing the f-stop and shutter speed combinations designed into the computer of a programmed camera. *Nikon, Inc.*

HAND-HELD LIGHT METERS

A light meter is a valuable accessory for the serious photographer, Figure 16-12. Even with cameras containing TTL light meters, mistakes can be made or difficult light measuring situations can occur.

There are two general categories of hand-held light meters: reflected and incident. *Reflected light* meter readings are taken with camera TTL light meters and with hand-held meters pointed

FIGURE 16-9. A brightly lighted background causes a TTL light meter to call for an f-stop and/or shutter speed that will underexpose the main subject.

FIGURE 16-11. An accurate close-up light meter reading gives proper exposure to the main subject.

FIGURE 16-10. Take a light meter reading close to the subject so the back lighting does not cause an inaccurate f-stop or shutter speed setting.

FIGURE 16-12. A hand-held light meter gives the photographer a wide variety of aperture and shutter combinations. *Quantum Instruments Inc.*

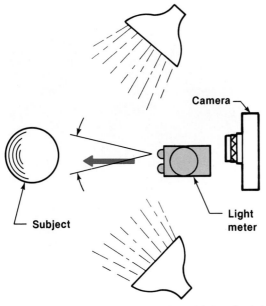

FIGURE 16-13. When measuring reflected light, the light meter is pointed toward the subject. *Quantum Instruments Inc.*

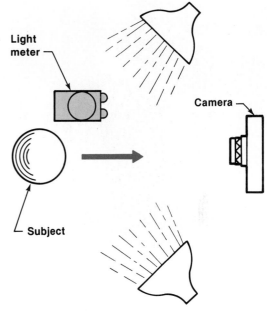

FIGURE 16-14. When measuring incident light, the light meter is held at the subject location and pointed toward the light sources. *Quantum Instruments Inc.*

toward the subject, Figure 16-13. *Incident light* meters measure the light falling on the subject, Figure 16-14. When using a light meter in the incident mode, a dome-shaped piece of plastic gathers light from 180 degrees, Figure 16-15. The microcomputer within the meter selects the most appropriate f-stop and shutter speed. It then provides the information on an LCD (liquid crystal display) or by moving a needle on a printed dial.

TYPES OF PHOTOCELLS

There are three major types of photocells, which react to light, that are used in light meters: *selenium*, cadmium sulphide (*CdS*), and silicon photodiode (*SPD*). Selenium is the most basic because it needs no batteries. It does not respond well to low light levels. CdS and SPD type meters must have batteries for electrical current. They are very sensitive, even for night and low light photography work. The majority of light meters in use today are of the CdS and SPD types.

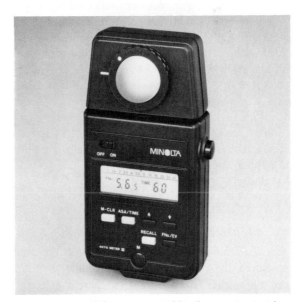

FIGURE 16-15. Light meters capable of measuring incident light have a distinctive dome-shaped white piece of plastic. *Minolta Corporation*

FIGURE 16-16. A spot meter is valuable when there is need to measure light in a very small area. *Minolta Corporation*

SPECIAL METERS AND ACCESSORIES

Special light meters and accessories help serious and professional photographers make valuable decisions. The *spot meter* measures reflected light and permits selective metering even at long distances, Figure 16-16. It also can be useful for taking pictures where a spotlight is used to light a performer on a stage. A spot meter covers only a very small area of 1 to 15 degrees; whereas, a regular reflective meter covers at least 40 degrees. Special adaptors permit readings from the viewing glass of view cameras, Figure 16-17. Many other special attachments have been designed to help the photographer measure light in those "hard-to-get" locations.

SUMMARY

To obtain good photographs, it is necessary to accurately measure light. Once the light has been measured, the correct aperture and shutter settings can be made. Through-the-lens camera metering and hand-held light meters are both valuable for the serious photographer. Automatic and programmed cameras are easy for nearly anyone to use.

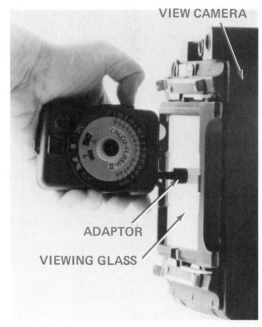

FIGURE 16-17. A special adaptor is used on a light meter to measure light falling on the viewing glass of a view camera. *Quantum Instruments Inc.*

REVIEW QUESTIONS

Answer these questions to test your knowledge of the unit content.

1. What do the initials "TTL" mean in photographic terms?
2. Photocells within a manually adjustable SLR camera are sometimes located:

   A. Behind the film
   B. Near the eyepiece
   C. Behind the mirror
   D. Inside the lens

3. Light striking an active photocell within a camera will cause the exposure meter needle to:

   A. Stay out of view
   B. Automatically center
   C. Swing out of control
   D. Move upward or downward

4. T/F Light meters within SLR cameras give accurate readings every time.

5. Over- and underexposures according to the light meter mean:

    A. The correct exposure may still be made
    B. Light meter is not working
    C. Too much or too little light will strike the film
    D. The battery is weak

6. An aperture priority camera is one that:

    A. Permits variable aperture openings
    B. Selects the correct aperture automatically
    C. Selects the correct shutter automatically
    D. Requires manual setting of the shutter

7. A programmed camera is designed:

    A. The same as an automatic camera
    B. To select the correct film speed
    C. For easier focusing than other cameras
    D. To select both the aperture and shutter

8. SLR cameras with manual override give:

    A. Limited flexibility for the photographer
    B. The photographer some direct control
    C. Some help for setting the shutter only
    D. Considerable light meter adjustments

9. T/F The hand-held spot meter permits readings from selected areas of a scene.

10. T/F The selenium type of photocell often is used for TTL light meters.

UNIT 17

SELECTING AND USING FILTERS

OBJECTIVES   *Upon completion of this unit, you will be able to:*
- *Explain the several characteristics of photographic filters.*
- *Organize filters into the basic categories.*
- *Select and use filters to obtain specific results for black-and-white and color photography.*
- *Handle and store filters in a proper manner.*

KEY TERMS   *The following new terms will be defined in this unit:*

Absorb	Filter Factor	Special Effects Filters
Color Compensating Filter	Neutral Density Filter	Transmit
Color Correcting Filter	Polarizing Filter	Ultraviolet Filter
Filter	Skylight Filter	Wratten

INTRODUCTION

The human eye has an amazing ability to adapt to different light sources. This is not true with photographic film, whether it be black-and-white or color. Film properties have been chemically formulated for average lighting conditions; thus, a change in lighting will cause the film emulsion to react differently. *Filters* can help correct this problem and improve the photograph. Many special effects also can be created with specially designed filters.

FILTER CHARACTERISTICS

Filters used in photography cause two important things to happen. First, a filter allows selected light to pass through. Thus, light is *transmitted* through the mass of material used to make the filter. Second, a filter allows selected light to be stopped from passing through. Thus, light is *absorbed* by the material used to make the filter.

    Light is transmitted and absorbed according to its color and the color or type of filter. White

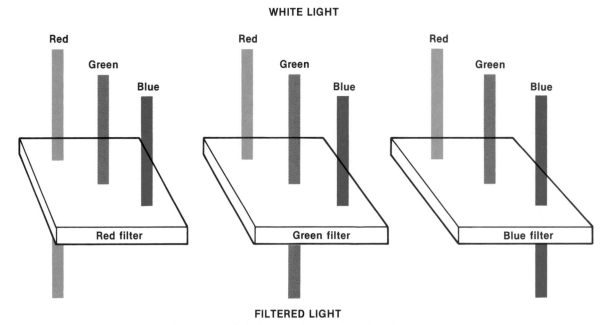

FIGURE 17-1. Filters transmit their own color and absorb their complementary colors.

light is made up of three primary colors: red, green, and blue. A colored filter will transmit its own color and absorb the other two colors known as complementary colors. For example, a red filter will transmit the red light waves and absorb the green and blue light waves, Figure 17-1. The same fact is true with filters of the other two primary light colors.

The density of the filter's color does make a difference in how much light of a given color is transmitted and absorbed. A light green filter will permit some red and blue light waves to be transmitted, while a very dark green filter will absorb some green light waves. A yellow filter transmits two of the primary colors, red and green, and absorbs the blue light waves, Figure 17-2. This happens because yellow light is created from a combination of red and green light waves.

Because most filters absorb some light, the photographer must remember to consider the *filter factor*. When a filter is used that has a factor of

more than 1, an exposure adjustment must be made. Using a filter with a factor of 2 means that twice the exposure will be needed. Actually, the filter is absorbing half of the light waves and transmitting the remaining half. Allowing more light to enter the camera and reach the film can be accomplished by opening the aperture, slowing the shutter speed, or a combination of the two. The correct exposure using a filter with a factor of 2 is achieved by increasing the aperture one full stop or decreasing the shutter speed by one full setting. Filters of different colors and densities have different factors; thus, it is important to be aware of these values, Figure 17-3.

It is easy to use filters with cameras that have through-the-lens (TTL) metering. Simply attach the filter to the front of the lens, take a light meter reading, and take the picture. Very likely, a larger f-stop or slower shutter speed will be indicated, but the photographer is certain that the correct exposure will be achieved. Whenever possible,

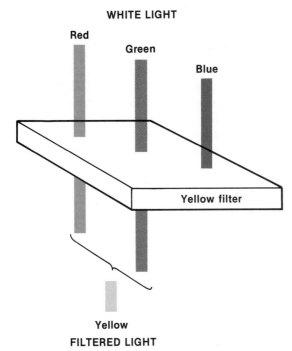

WHITE LIGHT

Red

Green

Blue

Yellow filter

Yellow

FILTERED LIGHT

FIGURE 17-2. A yellow filter transmits the two primary colors of red and green that combine to form yellow light.

Filter Factor	f-stop Increase
1.5	²/₃
2	1
2.5	1 ⅓
3	2 ⅔
4	2
5	2 ⅓
6	2 ⅔
8	3

FIGURE 17-3. Given filter factors require increases in aperture openings or the comparable reduction in shutter speeds.

use filters on cameras with TTL metering. It saves considerable time and ensures a higher level of success.

FILTER DESIGN

Filters are made of three basic materials: glass, gelatin, and plastic. Glass is rigid and is generally considered to be the best material for filters. Gelatin filters, often referred to as "gels," are thin and somewhat flexible. Plastic filters are tough and durable, but they do not have the optical qualities of glass or gelatin.

Glass filters are generally given a color by one of two methods. First, solid filter glass is made by adding the coloring to the glass in its raw molten state. This is done in large batches, making it difficult for precise control of color throughout the mixture. The laminated method is done by adding a thin layer of color between the two layers of clear optical glass forming the filter. The coloring is an integral part of the bonding material and can be applied with extreme accuracy of color and density. The bonding material has the same refraction characteristics as glass. Because of this, the filter made by this method is optically equivalent to quality solid glass.

Gelatin filters offer much variety because there is a great range of available colors. They are also quite economical when compared to either type of the basic glass filters. Unlike glass filters, they are affected by heat, making them susceptible to damage when being used in hot summer sunlight. Also, gelatin filters are easily scratched, making them very delicate to handle.

Plastic filters are economical, compared to glass filters. The optical characteristics are not equal to glass and gelatin, but they are very acceptable for many photographic uses. Several types and colors of plastic filters are available.

The two most common shapes of filters are round and square. Round filters made of glass are mounted in precision metal frames that either have threads or bayonet mountings, Figure 17-4. Both of

FIGURE 17-4. A set of high-quality glass filters mounted in metal bayonet mountings. *Mamiya America Corporation*

these mounting methods make it very easy to fasten one or more filters to the front of a camera lens. Most filters are placed in front of a lens, but some special lenses, such as fisheye and mirror telephoto types, take smaller filters mounted behind the lens.

Filter holder systems are designed for use on several kinds and types of camera lenses, Figure 17-5. These systems permit the use of square and round filters of either glass or gelatin. An adaptor ring must be obtained that fits the specific camera lens. The multiple position filter holder is held in place by the adaptor ring. Both round and square filters of either glass or gelatin are easy to install and remove in any combination desired.

FILTER IDENTIFICATION

Filters are identified by numbers, letters, and names. An 81A filter is light yellow in color. The number 81 refers to an identification system designed by an English company several years ago. Eastman Kodak purchased the company and utilized its *Wratten* filter numbering system. This system is considered the industry standard. The letter A on an 81A filter means that the yellow color is light or thin. An 81C indicates a yellow filter that is dark in color.

All companies do not use the same identification system. This makes it necessary to carefully study literature provided about filters and their results before selecting any for purchase.

FILTERS FOR BLACK-AND-WHITE FILM

Standard colors are used in producing filters that are useful in black-and-white photography. The typical colors are red, green, blue, yellow, and orange. These and other filters cannot add to the light falling on the film; they can only take away or absorb certain colors of light. It is important to remember that a filter transmits its own color and absorbs all or part of its complements.

Study the example shown, Figure 17-6. This provides proof that filters do make a significant

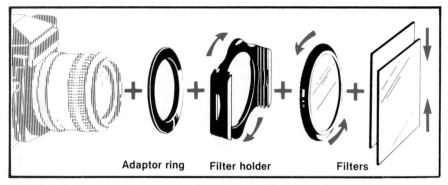

Adaptor ring   Filter holder   Filters

FIGURE 17-5. A universal filter system designed to fit different camera lenses and hold both round and square filters. *Cokin Creative Filter System, Minolta Corporation*

difference in black-and-white photography. The role of filters for black-and-white film is to correct picture contrast. Selecting the proper filter can render white clouds vividly bright against a dark sky or make the tones of colorful objects appear balanced and natural.

FILTERS FOR COLOR FILM

These filters are designed to improve the color match of light to the film. Light, whether it is artificial or natural, is not always the best color for the film in the camera. Film cannot adapt to different light sources as the human eye does. Thus, pictures taken with color film sometimes appear too yellow, red, or blue because of the color temperature (see unit 19) of the light source. Filter manufacturers and distributors have prepared brochures and booklets showing examples of pictures taken with various *color correcting* and *color compensating* (CC) *filters.* See unit 36 of this book for examples of pictures taken under various lighting conditions.

FILTERS FOR BOTH BLACK-AND-WHITE AND COLOR FILMS

Several filters are available that can be used with both black-and-white and color films. These filters are generally placed in four groups: ultraviolet, skylight, neutral density, and polarizing.

The *ultraviolet* (UV) *filter* is colorless but absorbs ultraviolet light waves that are always present in the atmosphere. Ultraviolet radiation can make distant landscape views appear abnormally blue. In black-and-white photographs, this radiation from the sun detracts from the overall contrast and detail. This filter does not cause an increase in exposure, but it does eliminate adverse effects caused by atmospheric haze. Often, the UV filter is called a "haze" filter. A wise photographer will leave a UV filter over the lens for regular photography. This helps protect the front element of the lens from dust and damage.

Skylight filters have a pink tinge of color. They are available in two or three densities and help to give better skin tones for outdoor portraits. They are also useful in reducing atmospheric haze.

NO FILTER

WITH YELLOW FILTER

FIGURE 17-6. Using a yellow filter enhances the contrast between the blue sky and white clouds in black-and-white photography.

Neutral Density Filters			
Number	**Density**	**Light Transmission Percentage**	**f-stop Increase Needed**
	0.30	50	1
ND 0.3	0.60	25	2
ND 0.9	0.90	13	3
ND 1.5	1.50	3.2	4 ½
ND 3.0	3.00	0.10	10

FIGURE 17-7. Neutral density filters absorb light and allow specific amounts to pass through.

Neutral density (ND) *filters* are used for exposure control. They reduce the amount of light passing through them without affecting the color of the light. There are various densities of ND filters. Each ND filter has a number that indicates the percentage of light transmission, Figure 17-7. This makes it possible for the photographer to have a wide selection of filters to lessen the amount of light entering the lens. ND filters can be used to reduce bright light conditions, making it possible to stop down the lens aperture or the shutter speed so a picture can actually be taken. This is sometimes needed with high speed film. Second, ND filters permit selective focusing for depth-of-field control. By using an ND filter that permits a small percentage of light to pass through, a larger aperture can be used. This reduces the "in-focus" area of the photograph.

Polarizing filters are designed to be rotated after they are fastened to the front of a camera lens. This permits the photographer to see the effect of the filter, assuming the camera is an SLR. In some cases, it is possible to hold a polarizer filter and look through it to determine the effects before attaching it to a lens. A polarizing filter is neutral in color and passes light waves that are vibrating in one particular plane, Figure 17-8. These filters are very helpful for the following situations:

- Color of blue skies is deepened for color photography.
- Contrast between the sky and clouds is increased in black-and-white photography.
- Reduction and removal of unwanted glare and reflections in shiny surfaces, such as glass and water, Figure 17-9. Does not work at certain angles and with metal surfaces.
- Helps to reduce haze in landscapes.

FILTERS FOR SPECIAL EFFECTS

These filters are available in a wide variety of colors, combinations, and shapes. Their purpose is to help produce special effects that give a wide range of emotional expression in photographs. One to three filters can be used at the same time to create an unusual photograph or one that enhances the scene. Booklets provided by filter manufacturers show many examples of using special effects filters to create interesting results for both black-and-white and color photography.

USING FILTERS

The best results will occur when filters are used on SLR cameras having built-in light meters. There is,

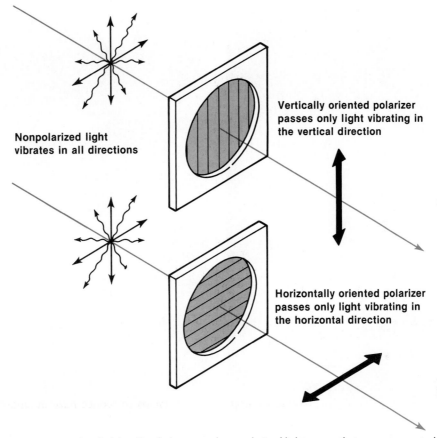

FIGURE 17-8.   A polarizing filter helps control nonpolarized light waves that cause unwanted reflections from shiny, nonmetallic surfaces.

though, no restriction in using filters on viewfinder, TLR, and view cameras. The main problem is with determining the exposure, but knowing and using filter factors permit quality pictures to be taken without undue problems.

The following points should be observed when handling and using filters:

- Never touch the filter surface with fingers as the acids in human skin can etch the gelatin and glass surfaces.
- Keep filters away from heat, moisture, and dirt. This is especially true with the less expensive gelatin filters.

- Clean filters with the manufacturer's recommended cleaning fluid and optical tissue.
- Replace filters in their cases when they are not in use.

SUMMARY

Filters are useful additions to a serious photographer's system of tools and equipment. A wise photographer will select filters with care. They should be obtained as situations arise or when a known lighting condition will exist. It is not necessary

WITHOUT FILTER

WITH FILTER

FIGURE 17-9.   A polarizing filter permits good photographs even with strong reflections from shiny surfaces such as this store window. *Kelvin K. Kramer*

to use filters for every picture; however, the UV filter serves as valuable protection for the lens. The polarizing filter is useful for both black-and-white and color photography; thus, it should be an early choice for a filter series.

REVIEW QUESTIONS

Answer these questions to test your knowledge of the unit content.

1. Filters cause light to be _____ and permit light to be _____ .

2. T/F Blue light waves pass through a red filter.
3. T/F A blue filter permits blue light waves to pass through its surface.
4. T/F Filter density has little to do with how light is affected.
5. When filters are used on the camera lens, exposures often must be adjusted. This filter characteristic is called:

   A. Filter adjustment
   B. Aperture control
   C. Filter factor
   D. Aperture/shutter factor

6. T/F Filters are easiest to use on cameras with TTL metering.
7. Filters are made of which three materials?
8. Which of the following is not used as a standard shape for filters?

   A. Oblong   C. Rectangle
   B. Round   D. Square

9. Filters are identified by _____, _____, and _____.
10. T/F The polarizing filter is useful with black-and-white and color films plus various lighting conditions.

—— UNIT 18 ——
COMPOSING PHOTOGRAPHS

OBJECTIVES   *Upon completion of this unit, you will be able to:*
* *Recognize photographs that have good composition.*
* *Use the eight composition guidelines to achieve pleasing arrangements of elements in photographs.*
* *Analyze a photograph and identify the composition guidelines utilized to achieve the given results.*

KEY TERMS   *The following new terms will be defined in this unit:*

Center of Interest	*Framing*	*Stop Action*
Composition	*Panning*	*Visual Perspective*
Converging Lines	*Photographic Eye*	

INTRODUCTION

It is important to remember that photographs are made and not simply just taken. The photographer must use the viewfinder of a camera to locate the best scene possible to record on film. To *compose* a photograph is one of the most important stages in the process of creating a photograph. Technical knowledge and ability plus elaborate equipment are of limited value unless the finished photograph is useful or is pleasant to look at.

COMPOSITION GUIDELINES

Photographic *composition* can be defined as, "a pleasing selection and arrangement of the elements within the picture." To achieve success with photographic composition is to please the viewers of the photograph, Figure 18-1. There are no hard-and-fast rules that a photographer must follow to obtain acceptable composition. There are, though, some

FIGURE 18-1. Photographic composition is in the eye of the beholder. *Sharon K. Fahey*

FIGURE 18-2. A photographer with the ability to see a photograph is a valuable person. *Marla Dory*

FIGURE 18-3. What interesting story can be told as a result of looking at this photograph? *Randy L. Slick*

guidelines that give strong direction to pleasing people who view and knowingly or unknowingly judge the photograph.

These suggested eight photographic composition guidelines can be used by both amateur and professional photographers. The quality or cost of the camera and accessories have no bearing on whether these guidelines improve the composition of the finished photograph.

- See a photograph before it is taken.
- Compose in the viewfinder.
- Create a center of interest.
- Use framing techniques.
- Divide scene into thirds.
- Observe background closely.
- Seek visual perspective.
- Be sensitive to motion.

SEEING A PHOTOGRAPH

Look around! Photographs are waiting to be made of the scenes that can be seen by the *photographic eye*. A photographer should be able to make useful photographs of nearly any visual image seen by the human eye, Figure 18-2.

Obviously, all photographs are not beautiful. Some are used strictly for information, such as those for science, technology, law enforcement, accident scenes, pathology, and others. Many photographs are used to record information that may not be pleasant to view. The photographer must be able to see what information should be recorded on film. This requires sensitive eyes of a skilled photographer. Once seen by the human eye, the camera and film are used to capture the image in a permanent fashion.

Pictures can be well planned in advance, or they can be unexpected and available in an instant. A photographer who can judge a scene quickly is a person capable of securing many useful and beautiful photographs. Keeping a photograph simple is one of the best "seeing" guidelines that any photographer can remember. Too much content in the recorded scene makes it difficult for the viewer to see the central theme of the photograph.

Photographs can be made to tell a story, Figure 18-3. Viewers should be able to look at a finished photographic print or slide and say something about what the image is showing. If nothing can be said, possibly the photograph should never have been made.

COMPOSE IN THE VIEWFINDER

The viewfinder in a camera is useful for more than just aiming the camera in the correct direction. It

FIGURE 18-4. The "hand" viewfinder often serves to quickly and accurately compose the scene.

FIGURE 18-5. The photographer was too far away, thus making it difficult to see detail in the subjects' faces.

FIGURE 18-6. Filling the viewfinder frame greatly improves the visible detail in a finished photograph.

can and should be used to carefully compose or arrange the scene content prior to release of the shutter. Film is wasted when the photographer fails to take even a few extra seconds of time to study the scene in the viewfinder.

A helpful technique in composing a photograph is to use the hands, middle fingers, and thumbs to form a rectangle. The "hand" rectangle can be held up to an eye and serve as a frame for selecting the best composition, Figure 18-4. This type of viewfinder gives considerable flexibility and saves time in selecting the best scene to capture on film.

Filling the viewfinder frame with the selected content is critical for clarity in the finished photograph. A common practice is to leave considerable space on all four sides of one or more people posing for a group picture, Figure 18-5. This makes the faces appear so small, making it difficult to distinguish significant detail. A much better practice is to move in closer and fill the viewfinder frame, Figure 18-6. Now detail is much clearer and viewers of the photograph know precisely who is being shown.

CREATE A CENTER OF INTEREST

Find something in the scene to focus the viewer's eyes upon, Figure 18-7. The photographer has the opportunity to select a portion of any picture and make it stand out. The angle the camera is held in relation to the main subject helps to determine how the subject will be viewed. A photograph with too many centers of interest is one with little or no interest at all.

FIGURE 18-7. A photograph that provides viewers with the opportunity to focus their attention has a good center of interest. *Rene' C. Gallet*

USE FRAMING TECHNIQUES

Artistic works such as paintings, needlepoint, and photographs are enhanced when mounted into frames. These wood, metal, or plastic frames draw attention to their contents. Some frames are better suited for selected artistic works than other frames. Creative talent is useful for good display of artistic work.

The same is true for content within the photographic scene. Architecture, landscapes, and seascapes can be highlighted when trees or manmade objects are used to create in-picture frames, Figure 18-8. A photographer needs to take time to look for natural *framing*. If this is not available, it is often valuable to create some type of frame. The goal posts on a football field can serve as excellent framing for educational activities—football team, band members, cheer-leading squad, and school friends.

DIVIDE SCENE INTO THIRDS

To help position the main subject with the photograph, it is useful to divide the rectangular area into

FIGURE 18-8. Natural framing with tree branches helps to focus attention directly on the main content of the photograph.

thirds, Figure 18-9. Divide the horizontal distance into three equal spaces with two vertical lines. Also, divide the vertical distance into three equal spaces with two horizontal lines. This gives four points of intersecting lines.

These points serve as guides to position the center of interest in the photograph. Any one of the four positioning points gives equal results. It is, though, important to consider the picture content while selecting the position point, Figure 18-10. The actual lines are not needed, because once this concept is known, it is easy to judge the location of the position points. The photographer should have little trouble in identifying these four points when looking through the camera viewfinder.

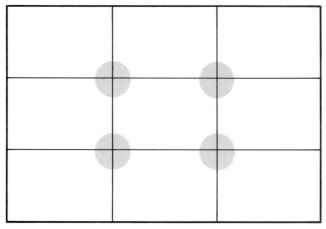

FIGURE 18-9. Dividing the picture area into thirds horizontally and vertically provides four positioning points for the center of interest.

FIGURE 18-10. The lower left position point works well for the center of interest in this photograph.

OBSERVE BACKGROUND CLOSELY

People appear to have objects spouting from their heads in some photographs, Figure 18-11. This can happen when the photographer fails to carefully check the background directly behind the subject. Other problems that can easily be recorded in a photograph show plants or flag poles growing out of people and animals. Look closely in the viewfinder for vertical objects that may cause abnormal backgrounds for the main content of any picture.

FIGURE 18-11. Unless care is taken, vertical background objects, such as stop signs, can appear to be growing from a person's head.

Horizontal type backgrounds often provide strange looking results too. With some thought, the photographer can move to a different location and take advantage of what is behind the center of interest. If possible, the subject can be moved forward, backward, or to either side, giving greater emphasis to the subject and less on the background.

Another problem centers around the edges of a photograph. Careful aiming of the camera eliminates problems of "cutting-off" a portion of the center of interest, Figure 18-12. Taking a few extra seconds to study the scene in the viewfinder can eliminate positioning problems near the picture edges. Intrusions in the picture area draw the viewer's eyes away from the main subject content. The unknown person and open door are unnecessary parts of the photograph shown in Figure 18-12.

SEEK VISUAL PERSPECTIVE

Perspective is important to consider in many aspects of picture taking. *Converging lines*, such as seen when looking down railroad tracks, give the viewer a sense of depth and distance. A desire to show a multi-story building as being tall and appearing to be reaching for the sky is easily achieved. Using a camera with a 50-mm lens, the photographer should stand near the building and aim the camera upward to fill the viewfinder with the entire building, Figure 18-13. Because it is farther away than the bottom, the top of the building recedes into the distance. It appears that if the lines in the building were to be extended, they would meet at a given point high in the sky.

The converging image problem can be reduced considerably by moving farther away from the building. Using a telephoto lens of 100 mm or more allows the photographer to fill the viewfinder with the selected content. The perspective of the building appears near normal when the camera is more equidistant from both the bottom and top of the structure.

FIGURE 18-12. Take time to aim the camera so that a part of the center of interest is not cut off as shown on the left side of this picture. Also, intrusions within the image area tend to draw the viewer's attention.

FIGURE 18-13. The vertical lines or edges of a tall building appear to converge when the photographer stands too close and tilts the camera up to get the whole building in the picture.

The photographer will do well to come down to the same level when taking pictures of little people, Figure 18-14. Children will react more favorably to the camera, and the camera lens will capture them in the best perspective. Something closer to the lens will look larger than it is in real life. Taking pictures from a normal adult height will make a child's head look larger and out of proportion to the rest of its body.

BE SENSITIVE TO MOTION

Moving objects can be photographed with precision. The two methods are called *stop-action* and *panning*. For stop-action, fast shutter speeds of 1/250 and above should be used when there is sufficient light and with fast films of ISO 200 and above. The results of using a fast shutter speed (to stop the action of a moving object) and a small aperture show all content of the photograph to be in focus, Figure 18-15.

Camera-to-subject distance and angle to each other make a significant difference in choosing shutter speeds. The line drawing in Figure 18-16 provides some guidelines when selecting a shutter speed to stop motion. There are many variables that must be considered whenever people or objects are moving within the picture area. Experimentation and bracketing are necessary for consistent and usable results.

Panning or moving the camera with the moving object gives interesting results, Figure 18-17. Two important techniques must be remembered. First, prefocus the lens on the spot where the picture will be taken. Second, keep the camera moving with the object before, during, and after squeezing the shutter release. Slower shutter speeds can be used with the panning method than with the stop-action method. Panning helps to focus the attention of the viewer on the center of interest, which improves the composition of the photograph.

FIGURE 18-14. To maintain good perspective, the photographer should get down to the same level as a small child. *Copyright 1991 David W. Coulter*

FIGURE 18-15. A racing bicyclist "frozen" on film while keeping the foreground and background in focus. Conditions: 50 mm lens, ISO 400 film, shutter 1/250, f-stop 5.6, and camera distance 12 feet (3.7 m). *Kelvin K. Kramer*

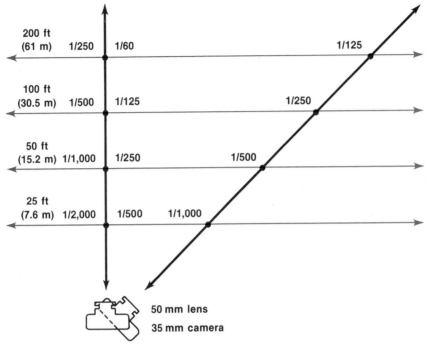

200 ft
(61 m)   1/250   1/60   1/125

100 ft
(30.5 m) 1/500   1/125   1/250

50 ft
(15.2 m) 1/1,000 1/250   1/500

25 ft
(7.6 m)   1/2,000 1/500   1/1,000

50 mm lens
35 mm camera

FIGURE 18-16. Shutter speed guidelines for stopping a person riding a bicycle. Colored lines represent 90° direction, and heavy black line represents 45° movement according to camera location.

FIGURE 18-17. Panning the camera with the moving object causes all stationary objects to be out of focus. A shutter speed of 1/90 was used for this action shot. *Kelvin K. Kramer*

SUMMARY

Composition guidelines should be considered each time a picture is taken. The eight guidelines highlighted in this unit are important to understand and use. In most situations, two or more guidelines are used in combination to obtain the desired photographic appearance. It is useful for the photographer to remember how each picture was obtained so the successful results can be repeated.

REVIEW QUESTIONS

Answer these questions to test your knowledge of the unit content.

1. Quality photographs are:

   A. Made only by professionals
   B. Created quickly
   C. Simply taken
   D. Planned and made

2. Photographic composition means:

    A. A specific style or kind of photograph
    B. A pleasing arrangement of picture elements
    C. Picture elements arranged in a specific way
    D. Material used to make the photographic print

3. A photographer needs a good photographic _____.

4. Why should the content of a photograph be kept simple?

5. T/F It is best to form the composition of the finished photograph in the viewfinder of the camera.

6. T/F A good photograph is one that has several centers of interest.

7. When should framing techniques be used by the photographer?

8. How many position points for locating the center of interest are created when the scene is divided into thirds both horizontally and vertically?

    A. 1   C. 3
    B. 2   D. 4

9. Taking a picture of a small child from a normal adult height will make the child's head appear larger or smaller (circle correct answer).

10. Following a moving object in the camera viewfinder is a useful technique called _____.

CHAPTER FOUR

LIGHTING THE SCENE

LIGHT FOR PHOTOGRAPHY

OBJECTIVES   *Upon completion of this unit, you will be able to:*
- *Discuss the science of light.*
- *Name and describe the characteristics of light.*
- *Classify light sources according to their color temperature.*
- *Explain how light is controlled in some areas of photography.*

KEY TERMS   *The following new terms will be defined in this unit:*

Camera–subject Axis	Inverse Square Law	Nanometer
Color Temperature	Kelvin Scale	Specular Light
Diffused Light	Light-color	Visible Spectrum
Electromagnetic Radiation	Light-direction	Wave
Electromagnetic Spectrum	Light-intensity	Wavelength
Flare	Light-quality	X-rays
Gamma Rays		

INTRODUCTION

Photography and light are synonymous, especially when recorded images made with cameras and film are discussed. All photographs must have a certain amount of light to create recorded images. Photographic film is more sensitive to light than is the human eye; thus, it is possible to capture images even when light cannot be seen. This is true primarily with film designed for the unseen *X-rays* and infrared light waves.

SCIENCE OF LIGHT

Light is energy, and all energy comes from the sun. Thus, the sun is the major source of light throughout every corner of the world. This has enabled photographic product manufacturers everywhere in the world to standardize their products. Also, different brands of photographic film can be interchanged in cameras, and predictable results can be obtained. Some characteristics of sunlight change, such as intensity and color, depending upon atmospheric conditions. Clouds, rain, and snow always alter the sunlight reaching the earth. Knowing how

to compensate for these difficulties permits the photographer to still accomplish the desired result.

Light travels at the enormous speed of 186,000 miles per second, Figure 19-1. This makes it seem instantaneous, but light emitted from the sun requires $8\frac{1}{2}$ minutes to reach the earth. Scien-

LIGHT, 7 1/2 TIMES AROUND IN ONE SECOND

FIGURE 19-1. Light travels at 186,000 miles per second, thus it can encircle the earth $7\frac{1}{2}$ times in 1 second.

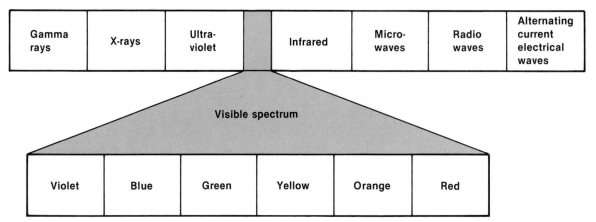

FIGURE 19-2. The electromagnetic spectrum includes the small but important visible spectrum so important to photographers.

tists speak of travel time in interstellar astronomy as light years. In a light year, a light beam will travel within a vacuum about 5,878,000,000,000 miles. For humans, the speed of light is so fast that little to no concern need be had about light needing time to travel from the photographic subject to the film.

The *electromagnetic spectrum* contains the entire range of wavelengths of *electromagnetic radiation* extending from *gamma rays* to the longest, radio waves, Figure 19-2. This full spectrum contains the narrow band of wavelengths known as the *visible spectrum* or visible light. The colors of violet, blue, green, yellow, orange, and red are observable in the visible spectrum.

Energy within the electromagnetic spectrum is transmitted in *waves*, Figure 19-3. *Wavelengths* of visible light are extremely small and are measured in *nanometers* (nm). One nanometer is equal to one millionth of a millimeter. The wavelengths of visible light extend from approximately 400 nm, which is seen as violet, to 700 nm, which is seen as red. The visible light colors when combined are seen as white light. To separate the individual colors, a prism can be used (see unit 36). Filters are used to transmit and absorb selected colors to enhance the quality of a photograph (see unit 17).

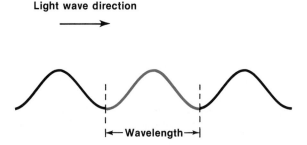

The short violet wavelengths are about 400nm long.

The longer red wavelengths are about 700nm long.

FIGURE 19-3. The length of a wavelength determines whether it is visible and, if so, its specific color.

CHARACTERISTICS OF LIGHT

There are four basic characteristics of light: intensity, color, quality, and direction. It is important that photographers consider not just one but all four of these characteristics.

Intensity refers to the amount of light falling on a subject. This characteristic does not determine whether a picture can be taken. It is relatively easy to measure light intensity with light meters; thus,

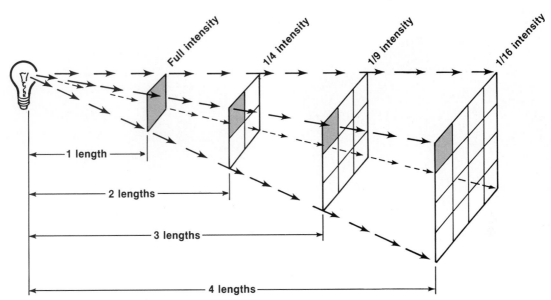

FIGURE 19-4. The effects of increasing the distance from the artificial light source is known as the inverse square law.

a difference in intensity helps to determine whether the exposures will be long or short, or whether a large or small aperture should be used. Low-intensity light can cause some problems when moving objects are involved in the picture.

The intensity of a single source of light falling on elements at different distances within a picture area does vary. This principle is known as the *inverse square law*. Simply stated, light intensity on close subjects is predictably greater than the light intensity falling on subjects farther away, Figure 19-4. For a given light intensity at one length of measure, the light intensity will be reduced to 1/4 when the length of measure is doubled. If the distance is increased to four times the original amount, the light intensity will be reduced to 1/16. Photographers must remember this important law when using artificial light in any way.

The *color* of light depends on the size of the wavelengths in the visible spectrum (see Figure 19-2). In most situations, the color of light has little effect on black-and-white photography. Most films for black-and-white photographs are capable of recording images illuminated by any or all of the colors in the visible spectrum.

Color film is very sensitive to the color of the lighting. Great care must be exercised to match color film with the color of the lighting (see unit 36).

The *quality* of the light refers to its harshness or softness. Harsh or *specular light* is that which comes from a spotlight or the sun on a clear, bright day. This type of light gives the subject a crisp and distinct appearance. The shadow edges will be sharply defined as well as dense black. This causes the contrasts between the lights and darks to be strong, Figure 19-5.

Softer or *diffused light* is available via the sun on overcast days. Also, special diffusers can be used with artificial lighting of all types. The shadow edges will be softer and less dense, Figure 19-6. Contrasts between the light and dark areas will be much lower than that produced by the harsh or specular light.

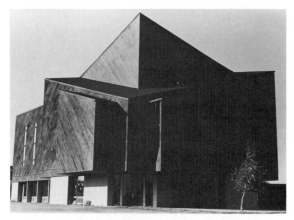

FIGURE 19-5. There are strong contrasts between the light and dark areas when harsh or specular light is used.

FIGURE 19-6. Softer, less harsh contrast is visible between the light and dark areas when diffused light is used.

The light characteristic of *direction* is very important for all photographers to consider. The direction the light is coming from in relation to the camera position greatly affects the appearance of the subject. The direction of light is given with reference to the *camera–subject axis*. This axis is an imaginary line drawn from the camera to the subject, Figure 19-7.

Light directed in front of the subject on the camera–subject axis is simply called frontal light. It is also called shadowless light because it casts no visible shadows from the camera lens angle of view. This type or direction of light is excellent for two-dimensional subjects, such as flat artwork. It has less value for three-dimensional (length, width, and depth) subjects because the light does a poor job showing the third dimension of depth.

Light on the camera–subject axis coming from behind the subject is called back light. This causes the subject to be lighted as a silhouette (see unit 23). Lighting of this type provides excellent dimensional shapes for length and width but does nothing for the third dimension of depth.

Ninety degree side lighting creates an interesting view and shows the surface texture of the subject very well, Figure 19-8. Too much contrast often is created; thus, supplemental frontal lighting

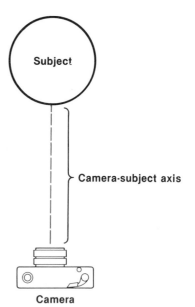

FIGURE 19-7. The camera–subject axis is an imaginary line drawn between the camera and the subject.

is frequently used to offset the harsh effects of side lighting.

A compromise to frontal, back, and side lighting is that which is simply called 45 degree lighting. Lighting from this angle gives good sur-

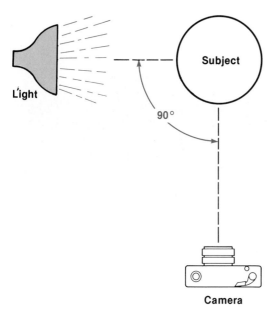

FIGURE 19-8. The results of 90 degree lighting give good surface texture and create high contrast among the many surfaces of the subject.

face texture and provides visual delineation of all three dimensions of the subject, Figure 19-9.

COLOR TEMPERATURE

Color temperature is one way to describe the quality of a light source. It is usually specified by a numerical figure from the *Kelvin scale*. This measuring system took its name from the English physicist, Lord William Thompson Kelvin, who designed the scale during the last half of the 19th century.

For use in photography, the scale is most often illustrated similar to a standard weather temperature thermometer, Figure 19-10. The major difference is that the various sources of light are shown at their degrees Kelvin (°K) on the scale. This helps the photographer quickly determine the °K of most light sources.

Color film is formulated by the manufacturer to react to light according to its color temperature. For example, daylight film is formulated (balanced) to give correct image colors for light that ranges from about 5,000 °K and upward. This includes sunlight, electronic flash and the once popular blue flashbulbs. Examples of color photographs taken under different °K lighting are shown in unit 36.

CONTROLLING LIGHT

Light can be and must be controlled for quality photographic results. Various filters are available that absorb some colors of light and transmit others. To balance the color of light to a given type of color film, color compensating filters can be used (see units 17 and 38). The intensity, color, or contrast of a given light source can be changed through filters for both black-and-white and color films for the purpose of increasing or decreasing elements of the picture taking scene.

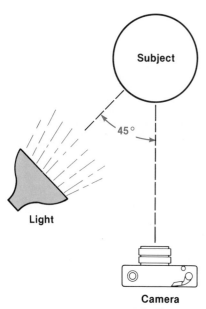

FIGURE 19-9.   Compromise lighting often referred to as 45° lighting.

Coatings on lenses used for cameras, enlargers, and projectors are placed there to protect the lens and to help control the light that passes through. Filters and film contain coatings on either or both sides so that the light waves will not stray or change color. Research is constantly underway by photographic equipment and supply manufacturers to improve on ways to control light generating equipment and light handling equipment, tools, and products.

Flare is a term describing uncontrolled light that passes through a lens and affects the image being recorded on the film. Lenses are multicoated so that bright spots do not occur on photographs or slides, Figure 19-11. It is difficult to see flare even when looking into the viewfinder of an SLR camera. The eye has the capability of compensating, but an uncoated or poorly coated lens does not. Flare can be controlled by using a lens shade that screws onto the front of a camera lens and, in some cases, a polarizing filter. Also, it is wise to avoid shooting into the sun or any other artificial light source. Sometimes flare can add to the photographic result, but this is usually considered to be an artistic happening.

SUMMARY

Light from the sun has been available since the beginning of time. The study of this form of energy began hundreds of years ago. In 1666, Sir Isaac Newton sent a beam of white light through a prism which broke the light waves into their separate colors. This was the beginning of what is known as the visible spectrum.

From the beginning, photography has been based upon using light to create recorded images. Early photographers knew little about light except to know that it took this important ingredient to

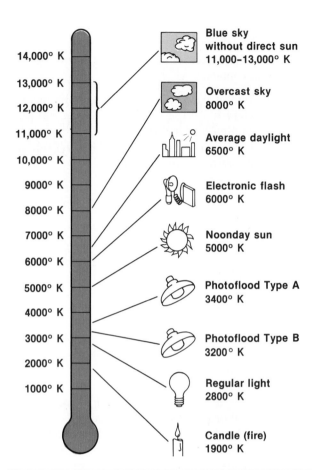

14,000° K — Blue sky without direct sun 11,000–13,000° K

13,000° K
12,000° K — Overcast sky 8000° K

11,000° K
10,000° K — Average daylight 6500° K

9000° K — Electronic flash 6000° K

8000° K

7000° K — Noonday sun 5000° K

6000° K

5000° K — Photoflood Type A 3400° K

4000° K

3000° K — Photoflood Type B 3200° K

2000° K

1000° K — Regular light 2800° K

Candle (fire) 1900° K

FIGURE 19-10. The Kelvin scale is used to illustrate the color temperature of various daylight and tungsten light sources.

VISIBLE FLARE   NO FLARE

FIGURE 19-11. Examples of photographs with and without flare.

FIGURE 19-12. Light made photography possible and early cameras gave people the chance to record images they saw. *Eastman Kodak Company.*

take pictures, Figure 19-12. Light is to photography as food is to life. Neither photography or life (plant and animal) can exist without the most basic ingredient.

REVIEW QUESTIONS

Answer these questions to test your knowledge of the unit content.

1. Why is light so important to photography?
2. Light is a major form of _____ that is emitted by the sun.
3. How many times can sunlight travel around the earth in one second?

   A. 3 times
   B. $7\frac{1}{2}$

   C. 9 times
   D. $12\frac{1}{2}$ times

4. How many minutes does it take light emitted by the sun to reach the earth?

   A. 2
   B. $3\frac{1}{2}$

   C. $5\frac{1}{2}$
   D. $8\frac{1}{2}$

5. T/F Light waves of 350 nm are easily seen by the human eyes.
6. Name the four basic characteristics of light.
7. T/F Knowing the principle of the inverse square law can be very helpful to a photographer.
8. Why is the direction of light important in relationship to the camera–subject axis?
9. When is the color temperature of light most important?
10. Name three standard methods of controlling light during picture taking with a camera.

DAYLIGHT PHOTOGRAPHY

OBJECTIVES *Upon completion of this unit, you will be able to:*
- *Analyze and effectively use sunlight throughout the day.*
- *Recognize creative photographs seen in shadows and patterns.*
- *Adjust aperture/shutter settings to take pictures in special weather conditions.*
- *Take quality pictures indoors using existing sunlight.*

KEY TERMS *The following new terms will be defined in this unit:*

Available Light	Diffused Light	Shadow
Baffle	Existing Light	Sunlight
Bracket Exposures	Patterns	

INTRODUCTION

Sunlight offers some of the best as well as some of the most difficult lighting conditions for a photographer. Sunlight can be intense or diffused, warm or cool, hot or cold, controlled or uncontrolled, friendly or unfriendly. Sunlight does wonders and because of it, an enormous number of photographic pictures are captured on film each and every day of the year. Using sunlight effectively is a skill that should be learned by amateur and professional photographers alike, Figure 20-1.

LIGHT THROUGHOUT THE DAY

Sunlight changes in intensity and color depending upon the time of day. Atmospheric conditions make significant differences in the quality of the sunlight that reaches the earth. Sunlight on any given day and time is different than any light that has ever fallen on a specific location before or in the future.

The difference is undetectable to humans and to many sophisticated instruments, but there is a difference. The atmospheric variables are impossible to accurately measure, and even the time of day is infinitely different from any other sun position and atmospheric condition that has ever existed. The comfort in this situation is that pictures taken in daylight do not have to be exactly the same as previous pictures. This is part of the excitement and challenge involved in taking pictures.

Some excellent photographs can be taken during the early minutes of dawn that begin each day, Figure 20-2. The low sun rays climb higher in the sky each passing second and reveal the mysteries of the night hours. As the morning wears on, the sun rises higher in the sky and the sunlight nears its normal color temperature. The sun no longer must shine obliquely through so much of the earth's

FIGURE 20-1. Sunlight was effectively used to highlight the focal point in this photograph. *Santa Fe Railway*

FIGURE 20-2. A patient photographer can obtain some prize-winning pictures during the few minutes of dawn that begin every day.

FIGURE 20-3. Mid-day sunlight creates few shadows and evenly lights the elements of the photographic scene. *Santa Fe Railway*

ever changing atmosphere. Thus, it gets closer and closer to what is commonly called natural sunlight.

Mid-day can be identified as the time from 11:00 a.m. to 1:00 p.m. Light during these two hours is approximately 5500 °K; thus, giving accurate colors, especially for color film. This is a good time to take pictures of buildings, equipment, and geographic locations, Figure 20-3. The straight or nearly straight overhead light creates few shadows and rather evenly illuminates the subjects.

Light during the afternoon and into the time period known as dusk offer excellent picture-taking opportunities. Subjects that for some reason do not look properly lighted during the morning hours may have the correct amount or angle of light during this portion of the daylight hours, Figure 20-4. Look for the right position and wait for the lighting that best accents the features of the subject.

CREATIVE USE OF SUNLIGHT

The absence or reduction in light to a given area that is surrounded by areas of high light levels is simply called *shadow*. Shadows are created, except for some very special lighting used for medical surgery, every time a dense object blocks light rays

FIGURE 20-4. Afternoon sunlight gives photographers many opportunities to select the best light for the subject being photographed. *Pete J. Porro, Jr.*

emitted from the sun or artificial sources. Conditions such as those illustrated in Figures 20-5 and 20-6 create some excellent photographic scenes.

Shadows are everywhere; thus, the photographer need only look. Upon locating a creative shadow image, it is useful to study the overall lighting and the various angles of view. Once selected, a light meter reading should be taken off the well-lighted area. A light meter reading taken in the shadow area would call for a larger aperture opening or slower shutter speed or both. This situation would then make the shadow area brighter, which is not the intended purpose.

Patterns, like shadows, are found in more locations than most people realize. It often takes the expert eye of a photographer to identify the interesting patterns that are both an act of nature and a creation of artistic people, Figures 20-7 and 20-8. Some very creative photographs can be obtained from pictures just waiting to be shot by photographers with perceptive eyes.

SPECIAL CONDITIONS

Shooting into the sun can cause flare and underexposure. To overcome the problem of flare, a lens hood can be used to shield the direct rays of the

FIGURE 20-6. Shadows can be found everywhere in natural settings.

FIGURE 20-5. Long shadows are created by objects that stand in the way of early morning sun's rays.

FIGURE 20-7. Natural patterns are some of the most interesting subjects for a photographer. *John Wolff*

sun. Sometimes, even this accessory will not protect the lens, and ultimately the film, from unwanted light. It may be necessary to use a large piece of

FIGURE 20-8. Patterns often are used in the construction and layout of human designed objects. *Richard Schneck*

opaque cardboard to block the direct sun rays from the lens, Figure 20-9. The *baffle* can be held by the photographer's assistant or hung by a string or wire from an overhead object. It is important not to have the baffle too low or it will block the angle of view of the lens.

Underexposure and even overexposure can easily occur due to false light meter readings. Examples of what can happen are shown, Figures 20-10 and 20-11. If a light meter reading is taken from the shadow areas, an overexposure will occur. The reverse is true if a reading is taken based upon direct sunlight. The solution is to reach a compromise, Figure 20-12. The photographer must make a judgment and establish an aperture/shutter setting somewhere between the extremes.

Pictures can be taken during poor weather. Fog, mist, haze, and even rain should not keep a photographer indoors. Keeping the camera and lens

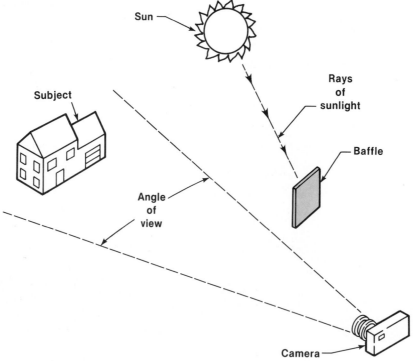

FIGURE 20-9. Flare can be eliminated by using an opaque baffle to shield sunlight from the lens.

dry is a problem that must be controlled. Much of the time it is possible to stand under a porch roof or within a building and take shots through open windows and doors.

Exposures in inclement weather can cause some difficulties. Fog, mist, haze, and rain often cause light meter readings to give underexposures. The moisture suspended in the air generally gives an overall bright tone to the scene. Because light meters built into cameras average the light, the reading is often inaccurate. For good results, it is best to open the aperture one to one and one-half stops, Figure 20-13. This guideline is not correct for all situations; thus, it is wise to bracket exposures when possible.

To *bracket exposures* means to take two or more exposures on either side of the exposure cal-

FIGURE 20-11. A print made from an underexposed negative. This is caused when the light meter is directed toward the sunlight.

FIGURE 20-10. A print made from an overexposed negative. This is caused when the light meter is directed toward the shadow areas of a scene that is backlighted by the sun.

FIGURE 20-12. This photograph illustrates a compromise in aperture/shutter settings between the under- and overexposures shown in Figures 20-10 and 20-11.

FIGURE 20-13.   Quality pictures can be taken during fog, mist, haze, or rain by opening the aperture 1 to 1-1/2 stops beyond light meter reading. *WW, Inc.*

culated as correct of the same scene. For example, the light meter indicates a shutter speed of 1/125 and an aperture of f/11. Leaving the shutter speed the same, an exposure would be made at f/16 and another at f/8. This gives three exposures, which helps to ensure that a correct exposure has been made.

Some excellent outdoor, daylight pictures can be taken during cloudy and overcast conditions. The clouds, especially when the cover is completely solid, serve as a diffuser to the strong and sometimes harsh sunlight. These conditions cause the light to surround the subject.

Snow can give problems when the photographer is determining the aperture/shutter settings. The brightness of the snow "fools" the light meter; therefore, it is necessary to compensate. A good way to determine the exposure is to take a light meter reading off the subject itself. The subject is always darker than the snow; thus, a more accurate reading will be given, Figure 20-14.

AVAILABLE INDOOR SUNLIGHT

Available light, sometimes called *existing light*, refers to the normal lighting found inside a building. Sunlight streaming into an indoor environment gives a photographer many opportunities to be creative. Light coming through windows casts both harsh

and soft shadows; thus, it is important to look for the best angle, Figure 20-14. The time of day makes a significant difference in the amount and angle of the light. The best lighting conditions from window sunlight can be determined by observing the light falling on the subject during an entire day. Assuming the sunlight is approximately the same the following day, take pictures at the best identified times.

Skylights used in the roofs of commercial and residential buildings provide excellent *diffused lighting*. Skylights made of "milky" looking plastic or glass are much better than those made of clear plastic or glass. They give excellent diffused lighting to indoor space just as an overcast day gives even lighting outdoors.

FIGURE 20-14.   Excellent photographs can be made of subjects with sunlight streaming indoors through a window. *The Morris Co.*

FIGURE 20-15. Pictures taken through windows from the center of a room offer striking contrasts.

Some unique pictures can be taken from the center of a room, Figure 20-15. Look through windows at the views found outdoors. Windows serve as frames and help to create panoramic views. Select the best indoor location and establish the aperture/shutter settings. Take a light meter reading from the camera location through the window. Also, take readings of the outdoor setting at the window location and again take a reading of a wall inside the room. Make certain the camera light meter does not get fooled by the direct window light. Compare the three exposure settings. If there are great differences, it will be necessary to determine a compromise exposure. It is also wise to bracket several shots and then select the best photograph after processing.

SUMMARY

An enormous number of pictures can be taken by using sunlight. Pictures are everywhere, both outdoors and indoors. Also, pictures are present during all types of weather conditions—rain, mist, haze, fog, and snow. Another important point to remember when taking pictures in sunlight involves measuring the available light. Light meters are fooled under certain conditions; thus, the photographer needs to compensate and override the aperture/shutter reading so that useful and creative photographs will result.

REVIEW QUESTIONS

Answer these questions to test your knowledge of the unit content.

1. T/F Photographs are always easy to take in sunlight.
2. Why is the sunlight falling on a specific location even at the same daily hour never the same again?
3. The normal color temperature of the mid-day sunlight is about:

    A. 2500 °K   C. 5500 °K
    B. 4000 °K   D. 8000 °K

4. T/F High-noon sunlight tends to cause few objectionable shadows.
5. The absence or reduction in light from a specific area is called _____.
6. Why do camera light meters sometimes give the wrong aperture/shutter information?
7. Shooting pictures directly into the sun can cause _____ and underexposure.
8. T/F Photographers should make a conscious effort to avoid taking pictures during fog, mist, haze, and rain.
9. The _____ sunlight caused during bright but overcast days is excellent for taking even lighted photographs.
10. T/F Good photographs can be taken indoors without the use of artificial lighting.

ARTIFICIAL LIGHT PHOTOGRAPHY

OBJECTIVES *Upon completion of this unit, you will be able to:*
- *Describe the several types of on-camera electronic flash units.*
- *Prepare the flash unit and camera for taking pictures.*
- *Take successful pictures with electronic flash equipment.*
- *Identify studio artificial lighting equipment.*

KEY TERMS *The following new terms will be defined in this unit:*

Artificial Light	*Flashbulbs*	*Macro Flash*
Dedicated Flash	*Flash Meter*	*Remote Flash Sensor*
Direct Flash	*Hot-Shoe*	*Sensing Cell*
Electronic Flash	*Indirect Flash*	*Strobe*
Fill Flash		

INTRODUCTION

Sunlight is not always dependable or available. Daylight is limited to a given number of hours per day depending on the time of year and location on earth. Many other variables affect the amount and quality of daylight; thus, it is important to have *artificial light* available for photography. Artificial light, whether it be incandescent bulbs, fluorescent tubes, flashbulbs, or electronic flash, can be con-

trolled by both amateur and professional photographers. This is important because no one likes to miss "getting a picture." Every aspect of the camera equipment, film, and lighting must work in perfect harmony.

ON-CAMERA FLASH

Flashbulbs in one form or another have been used for years. At one time, flashbulbs, flashcubes, and flashbars were the source of "extra" light for amateur as well as professional photographers. This type of instant lighting was very useful, but once the bulb had been used to create the light for the picture, it then had to be discarded.

Electronic flash units for 35-mm, SLR cameras are available in many sizes and models, Figure 21-1. A typical flash unit fastens to the bracket, called a shoe, on top of the camera, Figure 21-3. This places the flash head in direct alignment with the lens and illuminates the subject with direct lighting. Units of this type are activated with the shutter release through the electronic circuitry of the camera. The shoe contains circuitry of the camera. The shoe contains an electrical contact that matches with a contact point of the flash unit. This type of shoe is called a *hot-shoe.*

FIGURE 21-1. A special extension bracket was used to obtain this quality photograph. Extending the flash head several inches above the 35mm camera lens reduces or eliminates unwanted background shadows. *Siegelite Flash Bracket Co.*

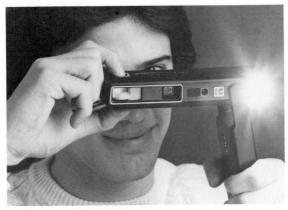

FIGURE 21-2. Electronic flash units are built into many viewfinder cameras. *Eastman Kodak Company*

Built-in *electronic flash* units are common on viewfinder cameras, Figure 21-2. The camera user need only point the camera at the subject and squeeze the shutter release. The flash can be used within a range of 4 to 11 feet (1.2 to 3.3 m) when using film with an ISO of 100. If ISO 400 film is used, the effective distance is extended to 16 feet (4.9 m). In-camera sensing determines whether flash is even needed to take a picture.

Another feature of the electronic flash illustrated in Figure 21-3 is the *sensing cell*. This unit measures the amount of light reflected from the subject. When the correct amount of light is provided to expose the film, the flash is shut off. This feature ensures a correct exposure everytime, assuming that all camera adjustments are made and the distance range of the flash is followed.

Larger electronic flash units have additional features, Figure 21-4. Units of this type permit wide flexibility and give the photographer greater opportunity to capture the desired scene on film. The sensing cell is adjustable depending on the camera–subject distance and the selected aperture. Also, the sensor can be set on partial power settings of 1/2, 1/4, or 1/16. The calculator dial is used to determine the sensor setting and lens f-stop depending upon the film speed and the camera–subject distance. The flash head is adjustable for use with wide

FIGURE 21-3. Sensing cells on electronic flash units turn the flash off when the correct amount of light reaches the subject. *Pentax Corporation*

CALCULATOR
DIAL

BATTERY
COMPARTMENT

VARIABLE
SENSOR

FIGURE 21-4. Larger flash units use variable sensors, calculator dials, and adjustable flash heads. *Vivitar Corporation*

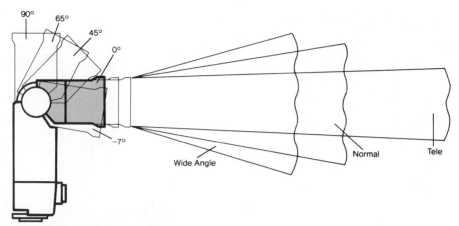

FIGURE 21-5. Some electronic flash units are adjustable for use with different lenses. Also, the head position can be changed for bounce lighting. *Vivitar Corporation*

angle, normal, and telephoto lenses, Figure 21-5. Also, the head is adjustable to different angles for bounce lighting.

Dedicated flash involves a marriage between the electronics of the camera and flash unit. The dedicated or integrated electronic flash system means that there is complete synchronization between the camera and flash. Integrated units permit the camera to control the flash duration by using the through-the-lens exposure information. Also, a change of aperture, film speed, lens, or the addition of filters over the lens is forwarded to the flash unit. The electronic circuitry in the camera and flash serves as a unified, computer-controlled system, Figure 21-6. With some dedicated flash units, a liquid crystal display (LCD) shows the photographer the information needed for a successful flash exposure. This information includes the speed of the film in the camera, the effective camera–subject distance range, the selected f-stop, and the flash head setting for the lens type (wide angle, normal or telephoto).

Electronic flash power sources vary from in-flash rechargeable nickel-cadmium cells to batteries of several sizes. Common among small and medium-sized flash units are the double-AA, 1.5-volt alkaline nonrechargeable and nickel-cadmium rechargeable batteries, Figure 21-7. Larger flash units that emit

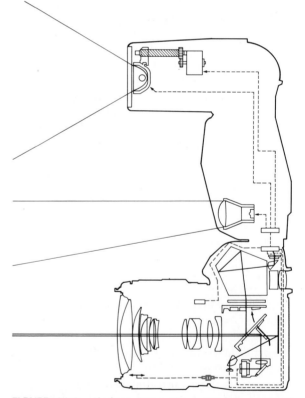

FIGURE 21-6. Flash emission is accurately metered through the lens and controlled by the electronic circuitry of a camera-dedicated flash system. *Minolta Corporation*

FIGURE 21-7. A recharging unit for size AA nickel-cadmium (Nicad) batteries. *Vivitar Corporation*

more light and have shorter recycle times use high-voltage battery packs, Figure 21-8. Units of this type are preferred by professional news photojournalists. The term *strobe* is sometimes used for the larger and more powerful electronic flash units. This is a misnomer, but its use dates back to the early days of electronic flash during the later 1940s.

USING ELECTRONIC FLASH

Electronic flash units vary in their specific use depending on their size and automation. The in-

FIGURE 21-8. A high-voltage flash system. *Pentax Corporation*

struction booklet that accompanies the flash or the camera should be used for specific information. The following steps need to be completed before using most flash units.

1. Install charged batteries in the flash unit. Make certain to match the polarity of each battery with the + and − markings in the battery chamber. If the flash unit has an "in-flash" rechargeable cell, make certain it is fully charged.

2. Attach the flash unit to the shoe on the camera. Make certain the flash unit is pushed all the way forward in the shoe so that the electrical contact points are in perfect alignment. Lock the flash unit in place.

3. Check the film ISO setting. Make certain that it is set on the film speed rating.

4. Set the exposure mode dial to the flash shutter speed. On most cameras, this setting is 1/60 or 1/125, Figure 21-9. Often there is a letter X or a bright-colored mark printed to the right of the correct shutter speed. It is necessary to use the correct shutter speed so that the shutter is open precisely when the flash lighting is at its brightest. Dedicated flash units control

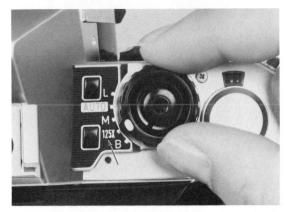

FIGURE 21-9. The exposure mode dial must be set for the correct shutter speed except for dedicated flash units.

the camera shutter speed when the flash power switch is turned on. The exposure mode dial must be set on the "A" (automatic) position.

5. Turn on the power switch and allow the batteries to charge the capacitor, which stores the high-voltage energy needed to produce the flash. This usually takes about 5 to 10 seconds if the batteries are strong.

6. Select the subject and focus the lens. Check the camera–subject distance on the lens barrel. Small flash units are useful for subjects at a maximum distance of about 13 feet (3.9 m) to 15 feet (4.6 m).

7. Set the aperture according to the information printed on or with the flash unit, Figure 21-10. The f-stop setting is determined by knowing the film speed and the camera–

subject distance. **Note:** Flash units with automatic sensors generally require a standard f-stop, and the sensor will turn the flash off when the proper amount of light has been emitted.

8. Take the picture. A second picture may be taken after the capacitor has been recharged and the ready light on the flash unit comes on. With dedicated flash units, it is possible to see the ready light in the viewfinder of the camera.

9. Turn the power switch off when finished taking pictures. Save the batteries from being drained of energy.

10. Remove the flash unit from the camera shoe. Do this carefully so that the shoe bracket does not become damaged on either the camera or the flash. Store the flash unit where it will be protected. Remove the batteries if the flash will not be used for one or more weeks.

TAKING PICTURES WITH FLASH

Electronic flash can be used to take pictures of large and small objects. Two or more flash units can be used together to provide the needed lighting. Remote flash sensors and extension flash cords can be attached to multiple flash units. This makes it possible for the photographer to activate all flash units at the same time by squeezing the shutter release.

DIRECT FLASH. Most flash pictures are taken by this method. With the flash unit in the camera or attached to the camera, the flash is pointed directly toward the subject. Upon squeezing the shutter release, the flash illuminates the subject instantaneously to expose the film. Place all subjects at least 6 feet (1.8 m) from a wall. This helps eliminate hard shadows being formed on the wall.

INDIRECT FLASH. Sometimes this is called "bounce flash." The light from the flash is deflected such that the strong light rays do not strike the subject in a direct manner. Three common methods of deflecting the flash lighting are shown, Figure 21-11. The first method is to point the flash head

FIGURE 21-10. Most flash units have printed information or an exposure dial to tell the photographer the correct aperture setting.

toward the ceiling. This bounces the light from the ceiling back down to the subject. A low, white ceiling is necessary for this method.

A second method uses a special diffuser glass or plastic sheet. This is positioned directly against the flash head. The diffuser causes the light rays to scatter widely. A third method involves a white card and a special card bracket. Pointing the flash straight up toward the 45° card causes the light to be deflected toward the subject.

All three methods of indirect flash are intended to soften the light reaching the subject. It causes the light to surround the subject, giving it depth and complete lighting.

FILL FLASH. Outdoor lighting does not always light the subject. Help from an electronic flash fills in the shadows, Figure 21-12. Set the flash unit on partial power, such as 1/2 or 1/4. Use the correct shutter speed for flash synchronization, and set the f-stop according to the average reading of the picture content. Take the picture in the usual manner. If possible, take two or three exposures at different settings (bracket). Translucent material can be placed over the flash head on units without power control. A clean, white handkerchief or tracing paper works very well.

MACRO FLASH. Using macro lenses to take pictures of small subjects at close range often de-

FIGURE 21-12. Fill flash provides light to shadow areas in outdoor pictures. *Top photo* was without flash, and *bottom photo* was taken with flash set at 1/2 power.

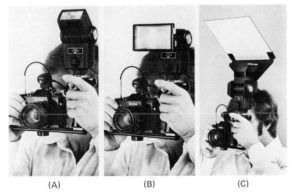

(A)   (B)   (C)

FIGURE 21-11. Three methods of deflecting the harsh rays of electronic flash. A. Pointing the flash head toward the ceiling. B. Using a special diffuser in front of the flash head. C. Attaching a white card above the flash. *Vivitar Corporation*

mands extra lighting. Round flash units called "ring-lights" can be fastened to the end of a lens. This provides light directly on the subject. *Macro flash sensors* can be used with flash units that have remote sensor capabilities, Figure 21-13. The special fiber-optic cord is fastened to the flash, and the sensor is attached to the front edge of the lens. A hot-shoe bracket that permits the entire flash unit to be

FIGURE 21-13. A fiberoptic sensor cord and a special hotshoe bracket permit flash lighting with macro lenses. *Vivitar Corporation*

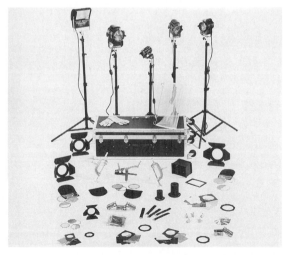

FIGURE 21-14. A wide variety of equipment is available for studio light distribution systems. *LTM Corporation of America*

tilted must be used. The flash head then can be directed toward the small subject.

STUDIO EQUIPMENT

There is a wide variety of studio artificial light distribution systems, Figure 21-14. These several systems contain incandescent and flash lamps, power packs, stands, and brackets. All equipment is designed and manufactured so that a studio photographer can control the lighting on the subject.

An important piece of lighting equipment for the professional photographer is an electronic flash system and umbrella, Figure 21-15. The amount of light output can be accurately controlled. Also, the umbrella diffuses the light, making it possible to

FIGURE 21-15. A multiple electronic flash and umbrella system for use in a professional studio. *Novatron of Dallas, Inc.*

surround the subject with light. Another useful device is the *flash meter*, Figure 21-16. Units of this type measure the electronic flash output and help the photographer determine the correct aperture and shutter settings.

Studio lamps should be fastened to sturdy stands, Figure 21-17. The best stands have strong wide legs and several telescoping poles. This permits the lamp to be positioned at several heights without fear of it falling over.

Another useful piece of studio equipment is a reflective shield, Figure 21-18. This device is useful in directing light to an exact location. It is possible to take advantage of unused artificial or natural light and direct the light back to the subject.

SUMMARY

Artificial lighting permits people to take pictures almost anywhere imaginable. Electronic flash is the most popular artificial lighting, but fluorescent and incandescent lighting are used too. The photographer must learn to control whatever form of artificial lighting is being used. Many electronic sensors and computers are being used in equipment to enhance lighting control.

FIGURE 21-17.   A strong and adjustable stand is necessary for a studio lamp. *Lowel-Light Mfg., Inc.*

FIGURE 21-16.   An electronic flash meter. *Vivitar Corporation*

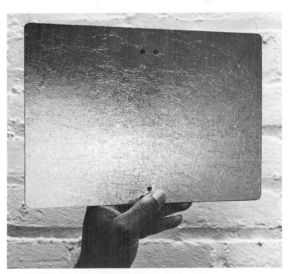

FIGURE 21-18.   A reflective shield is useful in directing light to specific areas. *Lowel-Light Mfg., Inc.*

REVIEW QUESTIONS

Answer these questions to test your knowledge of the unit content.

1. T/F Sunlight is a dependable source of light at all times of the day.
2. What source of artificial light was once very popular on cameras of all types and sizes?
3. T/F Built-in electronic flash systems are designed to flash when the sensing device determines a low light level.
4. The bracket containing electrical contacts on top of a camera that holds the electronic flash unit is called a _____.
5. The electronic flash sensing cell measures the light reflecting from the subject and:

   A. Turns off the flash unit.
   B. Tells the photographer the correct settings.

C. Changes the aperture of the lens.
D. Corrects the shutter speed.

6. A _____ flash unit is connected with the electronic circuitry of a camera.
7. Why do professional photographers generally prefer high-voltage electronic flash units?
8. T/F Batteries can be placed in an electronic flash unit with the contacts pointing in either direction.
9. Small electronic flash units are useful up to what camera–subject distance?

   A. About 10 feet (3.0 m)
   B. Less than 12 feet (3.6 m)
   C. Between 13 and 15 feet (3.9–4.6 m) maximum
   D. Greater than 20 feet (6.1 m)

10. Strong shadows are often caused when using the _____ flash method.

──────── UNIT 22 ────────

LIGHTING FOR PORTRAITS

OBJECTIVES *Upon completion of this unit, you will be able to:*
* *Assemble needed equipment for taking portraits and product photographs.*
* *Take studio portraits of acceptable quality.*
* *Capture the personality and characteristics of subjects in outdoor and environmental portraits.*
* *Use lighting sources to good advantage in product photography.*

KEY TERMS

Background Light	Hair Light	Rear Projection
Backgrounds	Key Light	Studio
Environmental Portrait	Outdoor Portrait	Studio Portrait
Fill Light	Portrait	Studio Props
Front Projection	Product Photography	

INTRODUCTION

People are interested in seeing photographs of themselves. They also want to see photographs of other people. *Portraits* are photographs of people taken, for the most part, in posed formal settings. Portraits can be taken in informal settings, too. The main difference between portraits and "snapshots" is that portraits are completely planned. Often portraits are taken of animals and birds, too, Figure 22-1.

FACILITIES AND EQUIPMENT

Photographic *studios* have many forms. Some are as large as airplane hangars, while others are as small as the hall walk-in closet. Studios are needed and used to take planned pictures for portraits and of products. A studio has one main purpose—to give the photographer control over the way a picture is made. This is done by protecting the subject or product from the weather elements. Also, a studio permits supports for various backgrounds and gives full control of the lighting.

   STUDIO SIZE AND SHAPE. Almost any rectangular room will work, but it should be at least 15 feet (4.6 m) wide. The length depends on the camera lens used, Figure 22-2. For a 50-mm lens, the length should be a minimum of 20 feet (6.1 m). A longer focal length lens of 80 mm to 135 mm is useful for portraits; thus, a greater camera—subject distance is required. It would be well to have a studio up to 30 feet (9.1 m) long to give the needed flexibility of camera and subject positions. The height should be

FIGURE 22-1.   An animal portrait requires much time, talent, and patience from the photographer. *Rulon E. Simmons*

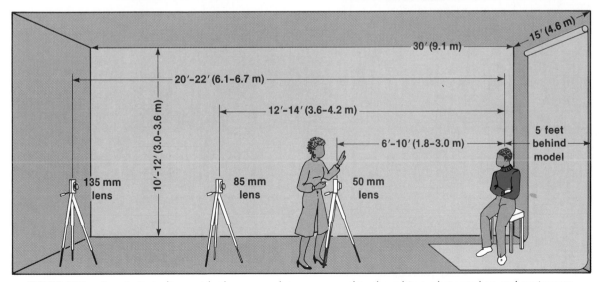

FIGURE 22-2.   A portrait studio must be large enough to accommodate the subject, photographer, and equipment.

a minimum of 10 feet (3.0 m). Higher ceilings permit greater flexibility with camera and lighting equipment.

LIGHTING. Lighting can vary between flash or tungsten (see unit 21). Both types of lighting have advantages and disadvantages. Tungsten equipment is more economical to purchase and is easier for the beginning portrait photographer to use. Electronic flash equipment is useful because it generates less heat and freezes the expression or action of the subject, Figure 22-3. Unless there will be continued use of the electronic flash lighting equipment, it is best to purchase and use tungsten lamps and stands.

Tungsten bulbs are available in many shapes and sizes. The standard photoflood bulb is available in 250 watt for a number 1 and 500 watt for a number 2.

BACKGROUNDS. Rolls of seamless background paper or window shade cloth are useful in creating attractive backgrounds. The rolls are fastened with a bracket to a wall near the ceiling and pulled down as needed like a window shade. Portable background stands that hold four different rolls are useful for a photographer who takes portraits in different locations, Figure 22-3.

Many different background scenes in full color are available. These can be hand painted, printed with a printing press, or created photographically on large rolls of photographic paper. Draperies designed for windows serve as excellent backgrounds. Rolls of cloth available from fabric stores also make useful backgrounds.

Backgrounds can be created using a special method called *front projection.* This system uses a combined SLR camera and projection unit, Figure 22-4. The basic principle involves using a 35-mm slide and projecting it toward the portrait subject. This is done in alignment with the lens of the camera, Figure 22-5. The projected background image completely surrounds the subject but does not affect the appearance of the subject. The background

FIGURE 22-3. A portable stand that holds several different backgrounds. *The Denny Mfg. Co., Inc.*

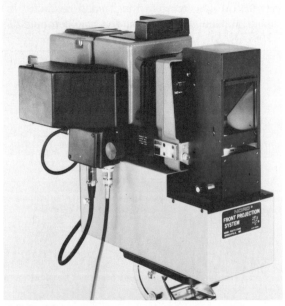

FIGURE 22-4. A front projection system consisting of an SLR camera and projector.

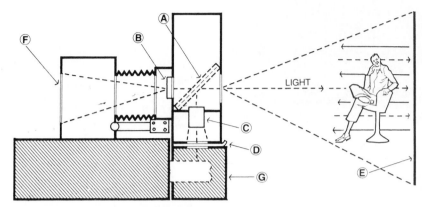

FIGURE 22-5. The key elements of a front projection system: *A*-beamsplitter, *B*-camera lens, *C*-projector lens, *D*-slide carrier, *E*-reflective background screen, *F*-SLR camera film plane, and *G*-slide illuminator.

screen has special reflective qualities that accept the projected image, making it look very natural, Figure 22-6. This system makes it possible for any background to be used where it is possible to take a slide in advance. *Rear projection* is also used to create portrait backgrounds.

STUDIO PROPS. Chairs, stools, benches, and tables are necessary props for portrait photography. Portrait photographers specializing in child photography must include toys, dolls, stuffed animals, and other interesting items. Beautiful outdoor background scenes sometimes demand outdoor-looking props, Figure 22-7. The list of studio props is endless. The creative portrait photographer will obtain and use the props best suited for the portrait subjects.

CAMERA. Portrait photography is best done with quality cameras, but nearly any camera can be used. SLR and view cameras are the best to use because greater control can be achieved. The popular 35-mm, SLR camera with a 50-mm lens gives acceptable results for amateur portrait photographers. A larger film size is better for portraits because the image does not need to be increased as much for portraits in the 8″ × 10″ and 11″ × 14″ size range. This makes the large format $2\frac{1}{4}$″ SLR

FIGURE 22-6. A portrait taken with the front projection system. Note the school library background used for the elementary age child. *Photo Control Corp.*

FIGURE 22-7. Outdoor-looking props are useful when using natural type portrait backgrounds. *The Denny Mfg. Co., Inc.*

camera and view cameras popular with professional portrait photographers.

STUDIO PORTRAITS

There are as many techniques and methods of arranging and taking studio portraits as there are portrait photographers; however, it is important to know the basic arrangement from which all other configurations result.

BASIC PORTRAIT ARRANGEMENT. The standard arrangement for studio portrait photography is shown in Figure 22-8. It contains a camera on a tripod, four flood lamps, chair or stool for the subject, and a background. The distance between the subject and the background should be 5 to 6 feet (1.5 to 1.8 m). The camera subject distance varies depending on the lens (see Figure 22-2). A 50-mm lens generally requires 6 to 8 feet (1.8–2.4 m) for a head to waist portrait.

SAFETY TIPS

1. Make certain the tripod and light stands are sturdy and fastened together properly. A stand could fall over and injure the subject or yourself.

2. Respect all electrical cords and connections. Electricity is very useful, but it can kill.

3. Keep the use of electrical cords to a minimum. Lift feet high so you or the subject will not trip and fall.

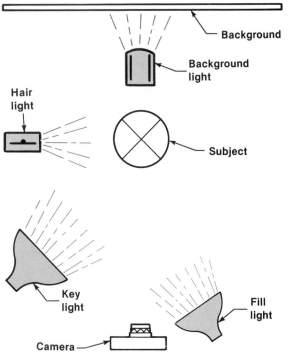

FIGURE 22-8. The basic arrangement for studio portrait photography.

4. Tell the subject to look away from the bright lamps from time to time. Severe eye fatigue can be caused by looking at bright lights for too long a time.

5. Do not touch hot light bulbs and equipment. Use a cloth towel to remove a hot but burned-out light bulb.

6. Make certain the background stand or roll of imaged paper/cloth is fastened securely. Never allow the background to fall and injure the portrait subject.

FLOOD LIGHTING PROCEDURE. Lighting arrangements vary depending upon the subject, photographer, studio, and conditions. The basic procedure for taking a portrait is as follows.

TAKING A PORTRAIT

1. Load the camera with film. It is best to use a slow film with an ISO rating of 100 or below. A slow film produces better enlargements because the silver halide light-sensitive particles are smaller than in faster film.

2. Set up the studio according to the basic arrangement shown in Figure 22-8. Make certain there are extra photoflood light bulbs and sufficient electrical cords and connections. Attach the camera to a sturdy tripod and fasten a cable release to the shutter release on the camera. A cable release is necessary so the picture can be taken when the subject is at his/her best.

3. Position the subject about 6 feet (1.8 m) from the background. The subject may be seated on a chair or stool. Sometimes it is best to have the subject stand without the use of any props.

4. Have the subject turn his/her body at a 30 to 45 degree angle to the camera—subject line. Then ask the subject to look back toward the camera lens. The arms and hands should be placed comfortably at the subject's side and lap.

5. Position the camera to include a head to waist viewfinder framing. Raise or lower the camera to match the subject's eye level. Focus the lens and set the shutter speed on 1/125. The f-stop will be set after the lights have been set into position.

6. Locate the *key light* on a line 45° to the left of the camera—subject line and about 2 to 3 feet (0.6–0.9 m) higher than the subject's head, Figure 22-9. The distance from the subject should be about the same as the camera equipped with a 50-mm lens.

7. Turn on the key light photoflood lamp. Move the light back and forth on the 45° line until highlights show on the forehead. Next raise or lower the light until a shadow from the nose touches the upper lip. Finally, move the key light left or right off the 45° line until the nose shadow merges with the cheek shadow.

8. Add the *fill light* by setting it as close to the right side of the camera as possible, Figure 22-9.

9. Turn on the fill light after adjusting the height to that of the camera. The fill light should be of less intensity than the key light. It should cast a slight shadow under the chin.

10. Place the *hair light* about 6 to 8 feet (1.8 to 2.4 m) high on the same side as the key light (see Figure 22-8). The reflector should be smaller than the one used for the key light. This gives a concentrated source of light.

11. Turn the hair light on. Adjust its position until a selected portion of the hair is highlighted, Figure 22-10. This area is generally very small and near the forehead.

12. Finally, set the *background light* between the subject and background (see Figure 22-8). The reflector should be rather large so as to light the entire background behind the subject.

13. Turn the background light on and position it as needed. Its intensity needs will vary depending on the color and light or darkness of the background material.

14. Take a light meter reading with all of the lamps on. Ask the model to hold a gray card so reflected light can be measured from it. Often these cards are called "18% gray cards" because they reflect about 18% of the light striking its surface. Gray cards serve as a standard reflective surface. They can be used in many photographic situations when light meter readings are needed. They work well for both black-and-white and color photography. A hand-held or a TTL camera light meter can be used. It may be necessary to remove the camera from the tripod and move up closer to the gray card. It's important to take the light meter reading only from the gray card.

15. Set the f-stop according to the light meter reading.

16. Take the first picture. Before doing so, talk with the subject and made certain he/she

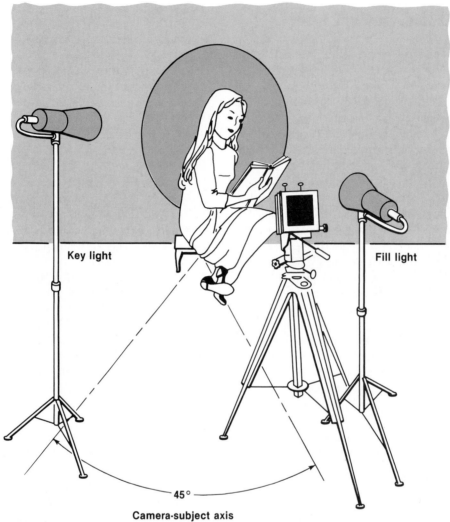

Camera-subject axis

FIGURE 22-9.   The positions of the key and fill lights in relation to the camera–subject axis.

is comfortable. Tell the subject the kind of expression wanted, and when it is seen, squeeze the shutter release.

17. Take 8 to 10 more exposures of the subject. Bracket the exposures, and seek different expressions from the subject.

18. Turn off all lamps and express appreciation to the subject for his/her cooperation. Explain

that proofs will be available in a few days.

19. Clean up and reorganize the studio. Make it ready for the next portrait sitting.

20. Process the film and prepare proofs for the subject to see. For black-and-white, see Chapters 5 and 6. For color, see Chapter 7. An example portrait of the basic four-light procedure is shown in Figure 22-11.

FIGURE 22-10.   The hair light creates a small bright area on the hair. *Photo Control Corp.*

FIGURE 22-11.   Pleasing portraits can be taken using the basic four-photoflood lamp procedure.

FLASH-UMBRELLA LIGHTING.   Professional portrait photographers often use lighting equipment of this type. Light placement is basically the same as that just described; however, there are many different lighting positions and techniques used in studio portraiture.

OUTDOOR PORTRAITS

Portraits taken outdoors must not be confused with snapshots. Both can be taken in the same location, but remember that portraits are fully planned down to the last detail. More attention is paid to the lighting, whether it be 100% sunlight or enhanced with artificial light.

Location is important to consider in outdoor portraiture. It is generally best to keep the subject(s) in familiar surroundings, Figure 22-12. The location

FIGURE 22-12.   A 50th wedding anniversary family outdoor portrait taken in the familiar backyard of the honored couple. *Clark's Photo Art Studio*

should help identify the people in the photograph. Familiar surroundings help to give the portrait a sense of authenticity.

Weather conditions play a big roll in outdoor portraiture. Too hot, too cold, rain, snow, or wind can be the cause of uncomfortable conditions. Flexibility and creativity are valuable assets to the outdoor portrait photographer.

Lighting is often best when the sun is the single source of light. Usually a mid-afternoon sun is suitable to give sufficient light to the subject's face. Appropriate shadows are also made possible. Reflectors and fill flash can be used to add needed lighting, especially for facial features. A key to quality outdoor portraits is to take light meter readings directly from the subject's face.

ENVIRONMENTAL PORTRAITS

People often like portraits taken in their own work surroundings. This type of setting is called an *environmental portrait*, Figure 22-13. It permits the photographer the opportunity to make a "comprehensive statement" about the person. As stated in unit 1, photography is a tool for communications. Thus, the environmental photograph is a way to tell more about a person than just show his/her image.

FIGURE 22-13. An environmental portrait communicates more about the subject than the standard studio portrait. *Birdie Kramer*

PRODUCT LIGHTING

In many respects, taking photographs of products is much like making portraits of people. Most lighting and equipment requirements are similar. Expressions and smiles are important with portraits but obviously are of no concern when taking pictures of inanimate objects. However, it is essential to show the product in the most favorable photograph that is possible.

There are many product photographs throughout this book. The lighting and background arrangements for many of these photographs were similar to that shown in Figure 22-9. Product lighting tables using formed, translucent plastic sheets permit the product to be surrounded by light, Figure 22-14. This lighting equipment helps to reduce shadow problems that sometimes occur in product photography.

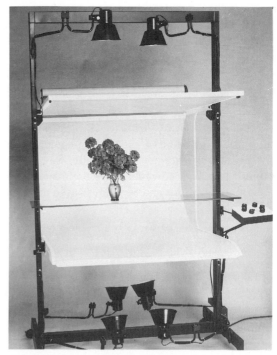

FIGURE 22-14. Background lighting tables with preformed translucent plastic permit products to be surrounded by light. *Leedal Inc.*

SUMMARY

Portraits are fully planned photographs of people and animals. Proper lighting of the subject is important to successful portrait photography. There are many methods and techniques used in portrait lighting. The four-light procedure is basic to understanding portrait lighting. Outdoor and environmental portraits can help to communicate the personality and characteristics of the subject. Product lighting is much like studio portrait lighting.

REVIEW QUESTIONS

Answer these questions to test your knowledge of the unit content.

1. T/F Portraits are photographs of people and products.
2. A portrait studio can be similar to a (an)

   A. Airplane hangar
   B. Living room
   C. Closet
   D. All of the selections

3. Name the two types of studio lighting.
4. T/F Backgrounds for portraits can be made from window drapery material.
5. The method of creating a background with a special camera–projector system is called _____.
6. Why are studio props valuable when taking portraits?
7. Identify the light that is placed next to the camera in the basic four-light portrait method.

   A. Key light   C. Hair light
   B. Fill light   D. Background light

8. T/F Hot photoflood lamps can be a safety hazard in a portrait studio.
9. The key light is often located at what degree from the camera–subject line?

   A. 45°   C. 90°
   B. 30°   D. 15°

10. T/F Snapshots and portraits taken outdoors often have the same quality.

NIGHT AND LOW-LIGHT PHOTOGRAPHY

OBJECTIVES   *Upon completion of this unit, you will be able to:*
* *Assemble and use accessory equipment needed for night and low-light photography.*
* *Compute exposures using specially prepared tables and calculators for night and low-light photography.*
* *Use existing light to obtain suitable photographs for night and low-light photography.*
* *Solve exposures and adjustments to film development involving reciprocity failure.*

KEY TERMS   *The following new terms will be defined in this unit:*

Exposure Calculator

Exposure Guide Table

Reciprocity Failure

Reciprocity Law

Silhouette Photograph

INTRODUCTION

Pictures can be taken at night and in low light, Figure 23-1. This is possible because there is enough natural or artificial light to expose film. Quality photographs and slides in both black-and-white and color can record sights beyond the vision of the human eye. It is, though, important for the photographer to use special techniques and utilize technical factors of light-sensitive film.

REQUIRED EQUIPMENT

Three pieces of equipment are essential for serious night and low-light photography. These are a tripod, a timer within the camera, and a cable release.

FIGURE 23-1.   The front entrance of a university library at night. *Barton A. Dennis*

FILLED WITH WATER

FIGURE 23-2.   Tripods can be made secure by fastening a weight in the center and letting it hang. *Kelvin K. Kramer*

TRIPOD. A tripod makes it possible to take exposures of 1/30 second up to several seconds. In some situations, exposures of several minutes need to be made. Several varieties of tripods are available (see unit 12). The heavy-duty type is best for long exposures because the camera must be perfectly still while the shutter is open. A light- to medium-duty tripod can be made secure by attaching a weight to it, Figure 23-2. A plastic pail and a small chain fastened directly under the camera

EXISTING-LIGHT EXPOSURE TABLE (AT ISO125)		
Picture Subject	**Shutter Speed (second)**	**Lens Opening**
In Homes at Night—Areas with Bright Light	1/30	f/2
Areas with Average Light	1/8*	f/2 ↓ 2.8
Candlelighted Close-ups	1/4*	f/2 ↓ 2.8
Indoor, Outdoor Christmas Lighting at Night	1*	f/5.6
Brightly Lighted Downtown Street Scenes at Night	1/30	f/2.8
Brightly Lighted Theater Districts—Las Vegas or Times Square	1/30	f/4
Neon Signs, Other Lighted Signs	1/60	f/4
Store Windows at Night	1/30	f/4
Floodlighted Buildings, Fountains, Monuments	1*	f/4
Distant View of City Skyline at Night	1*	f/2
Fairs, Amusement Parks	1/30	f/2
Aerial Fireworks Displays—Keep camera shutter open for several bursts	BULB or TIME*	f/11
Night Football, Baseball, Racetracks†	1/60	f/2.8
Basketball, Hockey, Bowling	1/60	f/2
Boxing, Wrestling	1/125	f/2
Stage Shows—Average Lighting	1/30	f/2.8
Bright Lighting	1/60	f/4
Circuses—Floodlighted Acts	1/30	f/2.8
Ice Shows—Floodlighted Acts	1/60	f/2.8
School—Stage and Auditorium	1/15*	f/2
Swimming Pool—Indoors, Tungsten Lights Above Water	1/30	f/2
Church Interiors—Tungsten Lights	1/15*	f/2

*Use a tripod or other firm camera support. †When lighting at these events is provided by mercury-vapor lamps, you'll get better results by using daylight-type film.

FIGURE 23-3. An exposure guide table is useful for night and low-light photography. *Eastman Kodak Company*

serve well. The pail can be filled with water, sand, or other heavy material. Exercise care as objects or material that are too heavy may damage the tripod.

TIMER WITHIN CAMERA. Low-light pictures can be taken conveniently with a timer on the shutter. Actually, all adjustable cameras have a timer on the shutter. A shutter speed of 60 is equal to 1/60th of a second; thus, a timing device is necessary. It is helpful to have cameras that allow camera-controlled shutter speeds of 1, 2, 3, and 4 seconds.

CABLE RELEASE. A cable release with a set-screw is the most useful for exposures longer than the 4-second camera-controlled timer exposures (see unit 12). To use, position the camera exposure mode on "B." Establish the exposure length, and fasten the cable release to the shutter release button. Press the cable shaft to open the shutter, and tighten the set-screw. When the exposure time is nearly up, hold in on the cable shaft and loosen the set-screw. Release the shaft slowly so that a sudden movement does not vibrate the camera.

MEASURING LOW LIGHT

Light meters within a camera and hand-held models can be used to measure many low-light conditions. Care must be taken so that the light meter is not misled by a single bright light within the scene.

A light meter, hand-held or within the camera, that cannot measure the low light at a given film speed setting can be readjusted. For example, with the film speed set at ISO 100, the meter will not give a reading. Move the film speed setting to 200 and take a reading. If the light meter still does not react, move the film speed to 400. At this setting, the exposure length reads 1 second. Each increase in film speed is equal to one full f-stop or shutter speed; thus, the exposure for ISO 400 is 1 second, ISO 200 would be 2 seconds, and ISO 100 would be 4 seconds.

EXPOSURE GUIDE TABLE. Information from these sources is based upon experience of the film manufacturer. Guide tables are sometimes available with rolls of film, Figure 23-3. Existing light expo-

sure tables of this type often provide aperture and shutter settings based upon a given film speed rating. For different film speeds, it is necessary to make adjustments in either the aperture or shutter settings or both.

EXPOSURE CALCULATORS. These devices are very helpful because various film speeds and types often are built into the system, Figure 23-4. Exposure calculator information is based upon trial and error just as exposure guide tables are. To use the calculator shown in Figure 23-4, complete these steps:

USING AN EXPOSURE CALCULATOR

1. Select the scene number according to a listing of existing light scenes, Figure 23-5.

2. Note the film type. Using the suggested film type will give better color reproduction with color films. Black-and-white film can be used with any scene lighting.

3. Match the scene number with the film speed by turning the inside dial, Figure 23-6. Example: Scene number 13 is lighted Christmas decorations both indoors and outdoors. Assume a film speed of ISO 64.

4. Select an f-stop to correspond with the needed or desired depth of field. Example: f/2.8–f/4.

5. Determine the shutter speed or time exposure. Example: f/2.8–f/4 = 1/2 second exposure.

6. Set up the camera, make the needed adjustments, and take the picture. As with all night and low-light photography, it is best to bracket the exposure.

NATURAL EXISTING LIGHT

Excellent pictures can be made using light from the rising sun and setting sun, Figure 23-7. Pictures can

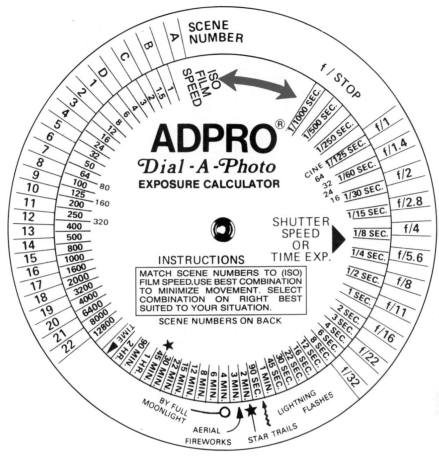

FIGURE 23-4. An exposure calculator that provides aperture and shutter settings based upon various film speeds and lighting conditions. *Dial-A-Photo System*

be taken during most of the nighttime hours from dusk to dawn. Moonlight is an excellent source of light, especially for landscape scenes. A typical exposure of a landscape lighted with a full moon is 4 minutes at f/8 with ISO 125 film, Figure 23-8. If the landscape is covered with a blanket of snow, the exposure will remain at 4 minutes, but at f/22. The reduced aperture is possible because the snow makes the landscape much brighter.

Silhouette photographs can give a dramatic effect and are relatively easy to take, Figure 23-9. The important thing to remember is to keep the subject between the light source and the camera. Take a light meter reading from the light source itself and not the subject.

ARTIFICIAL LIGHT

Various artificial light sources can be used for night-time pictures, Figure 23-10. Incandescent bulbs, fluorescent tubes, and quartz lights are common in industrial, business, and home settings. Film choice is important when using color film. For the best results, use a tungsten film. This will give the best

SCENE NUMBER	DAYLIGHT and EXISTING LIGHT SCENES (D) USE DAYLIGHT FILM (T) USE TUNGSTEN FILM (A) USE ANY FILM	FILM TYPE
(A)	BRIGHT SUN (AVERAGE SUBJECT). CLOSE DOWN 1 f/STOP ON SAND OR SNOW.	(D)
(B)	BRIGHT SUN SIDE LIGHTED CLOSE UPS.	(D)
(C)	CLOUDY BRIGHT (NO SHADOWS), SUNSETS.	(D)
(D)	HEAVY OVERCAST, OPEN SHADE.	(D)
(1)	ICE SHOWS WITH WHITE SPOTLIGHTS.	(D)
(2)	OPEN SKY WINDOW LIGHTED SUBJECTS.	(D)
(3)	BRIGHT STAGE AND THEATER SHOWS,	(T)
(4)	CIRCUS (SPOTLIGHTED ACTS), ICE SHOWS (COLORED LIGHTS), NEON AND OTHER LIGHTED SIGNS, MOVIE MARQUEES.	
(5)	SKYLINE 10 MINUTES AFTER SUNSET.	(D)
(6)	TIMES SQUARE, LAS VEGAS STREET SCENES, ELECTRIC LIGHTED PARADES. NIGHT FOOTBALL, BASEBALL, BOXING, RACE-TRACK, CAMPFIRES, BURNING BUILDINGS, STORE WINDOWS, BLACK & WHITE TV AT 1/25 SEC. FIREWORK DISPLAYS ON GROUND.	(T)
(7)	BRIGHT FLUORESCENT LIGHTED STORES, OFFICES AND RESTAURANTS, LIGHTED DISPLAY CASES. CIRCUS (GROUND LEVEL ACTIVITY). BRIGHT HOME INTERIOR.	* (T)
(8)	DOWNTOWN BRIGHT STREETS (TRY WET STREETS FOR REFLECTIONS). USA MAIN STREET. BASKETBALL, HOCKEY, BOWLING. AVERAGE LIGHTED STAGE SHOWS.	(T)
(9)	HOSPITAL NURSERY, SWIMMING POOLS, SHOPPING CENTERS.	(T)
(10)	AMUSEMENT PARKS, CARNIVALS, FAIRS. RIDES USE LONG EXPOSURE TIME FOR INTRICATE DESIGNS. VEHICLE TRAFFIC	(T)
(11)	COLOR TV AT 1/15 SEC. (D). BOAT, HOME AND FLOWER SHOWS, HOME INTERIORS, GAS STATIONS.	(D) and (T)
(12)	CHURCH INTERIORS, SCHOOL STAGES, AIRPORTS, BUS, TRAIN TERMINAL, HOTEL AND HOSPITAL LOBBIES.	(T)
(13)	LIGHTED CHRISTMAS DECORATIONS CHRISTMAS TREES (IN AND OUTDOORS).	(A)
(14)	AIRLINE, BUS, TRAIN INTERIORS.	(T)
(15)	CANDLELIGHT (CLOSE-UPS). FLOOD-LIGHTED FOUNTAINS, BUILDINGS, MONUMENTS.	(T)
(16)	INDUSTRIAL PLANTS (OVERALL VIEW). STREET CORNER LIGHTING.	(A)
(17)	ILLUMINATION UNDER STREETLAMPS OF SIDESTREETS.	(A)
(18)	DIM NIGHT CLUBS, DANCE FLOORS AND BANDSTANDS.	(T)
(19)	NIAGARA FALLS (WHITE LIGHTS). OPEN 1 f/STOP FOR COLORED LIGHTS.	(A)
(20)	SKYLINE OF BUILDING, BRIDGES, DISTANT VIEWS WITH SCATTERED LIGHT.	(A)
(21)	BOATYARDS, DOCKS, WHARVES.	(A)
(22)	MOONLIT LANDSCAPES (SNOW ON GROUND) DO NOT INCLUDE MOON. WITHOUT SNOW OPEN UP 2 TO 3 f/STOPS.	
•	GLASSWARE IN WINDOW DISPLAY. TAKE FROM INSIDE. USE 1 f/STOP MORE THAN DAYLIGHT CALLS FOR.	(D)
•	CHURCH STAIN GLASS WINDOWS. DAYTIME FROM INSIDE. USE 3 f/STOPS MORE THAN OUTDOOR LIGHT CALLS FOR.	(D)
•	TUNGSTEN FILM CAN BE USED IN DAYLIGHT WITH AN 85 B FILTER.	
*	USE FLUORESCENT FILTER OPEN 1 f/STOP.	
O	SCENES LIGHTED BY FULL MOON.	(D)

FIGURE 23-5. A listing of regular and low-light scenes including a recommended type of film to use. *Dial-A-Photo System*

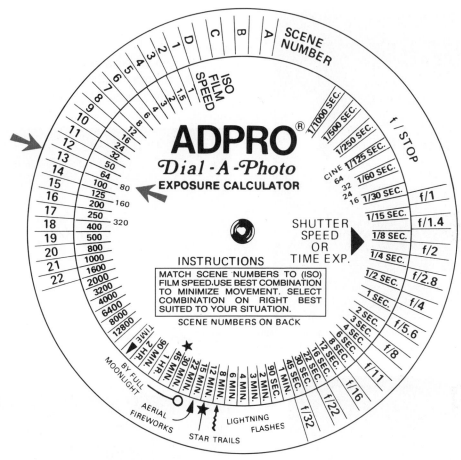

FIGURE 23-6. Matching the scene number with the film speed gives several f-stop and shutter speed combinations. *Dial-A-Photo System*

color in most artificial light environments. Black-and-white film can be used under any type of lighting and color temperature.

Moving light sources, such as automobile headlights and taillights, are enjoyable to capture on film. Another source of exciting photography is fireworks. A typical set-up would include color film, tripod, cable release, and an unrestricted view of the fireworks area. Using ISO 400 film and f/2.8, the exposure would be about 4 minutes. This would very likely include several bursts of fireworks.

RECIPROCITY LAW

The *reciprocity law* can be stated in mathematical terms as,

exposure = intensity of light × time or $E = I \times T$.

This simply means that the various apertures and shutter speeds can be used in different combinations to expose film to the same amount of light, Figure 23-11. Increasing the aperture one full f/stop doubles the amount of light striking the film. By

FIGURE 23-7. The low setting sun provides scene highlights not available during the full daylight. *Santa Fe Railway*

increasing the shutter speed by one full step, the amount of light striking the film is back to the original amount.

Reciprocity failure occurs when the film emulsion does not react to the stated scientific reciprocity law. Failure occurs when the shutter speed is

FIGURE 23-9. The silhouette of the tree branches makes them stand out against the frost-covered foliage. *Richard L. Johnson*

FIGURE 23-8. Landscape lighted by a full moon—time exposure of 4 minutes, f/8, ISO 125 film. *Kelvin K. Kramer*

FIGURE 23-10. Artificial lights provide sufficient illumination for many nighttime pictures. *Santa Fe Railway*

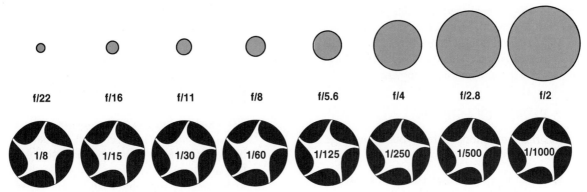

FIGURE 23-11. Each of these f-stop and shutter speed combinations admits an equal amount of light through the lens. This is defined as the reciprocity law.

shorter than 1/1000 second and as long as 1 second and longer. The problem occurs because of the way that individual silver halide crystals in the film emulsion are exposed.

Light is made of millions of tiny particles known as photons. When the camera shutter is open, the photons pass through and strike the film. A minimum of four photons must strike a silver halide crystal at the same time to make it exposed. Extremely fast shutter speeds do not permit enough light to pass through for the photons to do their work. The same is true for long exposures in low light.

It is important to compensate for the failure of the film emulsion to become exposed. The solution is to provide more time for the light (photons) to strike the film. Also, it is necessary to adjust the film development time to compensate for the expected exposure change. Charts and tables have been prepared to assist the photographer with the condition known as reciprocity failure, Figure 23-12.

SUMMARY

Night and low-light photography can be challenging but rewarding, Figure 23-13. A good tripod,

Indicated Exposure Time	USE		Development Time Change
	EITHER This Aperture Change	OR This Adjusted Exposure Time	
1/100,000 sec	+ 1 stop	Use Aperture Change	+ 20%
1/10,000 sec	+ ½ stop	Use Aperture Change	+ 15%
1/1,000 sec	None	No Adjustment	+ 10%
1/100 sec	None	No Adjustment	None
1/10 sec	None	No Adjustment	None
1 sec	+ 1 stop	2 sec	− 10%
10 sec	+ 2 stops	50 sec	− 20%
100 sec	+ 3 stops	1200 sec	− 30%

FIGURE 23-12. Reciprocity failure requires an increase in exposure and adjustments to the film development. *Eastman Kodak Company*

timer in the camera and possibly a wristwatch with a second hand, and a cable release are almost essential for good results. Exposure guides and calculators are a must, too. Finally, reciprocity failure requires film exposure and development adjustments.

REVIEW QUESTIONS

Answer these questions to test your knowledge of the unit content.

1. Name three pieces of equipment that are essential for night and low-light photography.
2. How can a light-duty tripod be made more firm and secure?
3. T/F All adjustable 35-mm cameras contain a timer within the shutter mechanism.
4. What mode should the camera be set on when a time exposure longer than 4 seconds will be taken?

   A. Program   C. Auto
   B. Manual   D. "B"

5. T/F Light meters are ineffective in low light.
6. Study the exposure calculator in Figure 23-6. Would it be possible to take a picture hand-holding the camera with the shown shutter–aperture combinations? Why?
7. T/F Existing light exposure charts and calculators are accurate enough so that bracketing can be considered a waste of film and time.
8. Exposures for moonlight landscape scenes must be adjusted when _____ is on the ground.
9. What type of picture is taken when the subject is located between the camera and source of light?
10. Reciprocity failure does *not* occur with which of these film exposures?

    A. 1/1000 second   C. 10 seconds
    B. 1/100 second   D. 100 seconds

FIGURE 23-13. Quality night photographs can be obtained through cooperation of the subjects plus patience and skill of the photographer. *The Morris Co.*

FIVE

FILM PROCESSING—BLACK-AND-WHITE

CONTINUOUS-TONE FILM AND IMAGING

OBJECTIVES *Upon completion of this unit, you will be able to:*
- *Prepare a sketch showing the six layers of black-and-white photographic film.*
- *Explain the six common film characteristics.*
- *Tell the meaning of latent image and value of the characteristic curve.*
- *Compare current and early film technology.*

KEY TERMS *The following new terms will be defined in this unit:*

Acutance	Emulsion	Latent Image
Bar Code	Gelatin	Orthochromatic
Blue-Sensitive Film	Grain	Panchromatic
Characteristic Curve	Halation	Silver Halides
Checker Code	High Contrast Film	Spectrogram
Continuous-tone Film	Infrared Film	

INTRODUCTION

Continuous-tone film is the type used in cameras throughout the world. The film is capable of recording images from very light to very dark. It is so sensitive to light that most images, light or dark and everything in between, seen by the human eye can be captured on film.

FILM STRUCTURE

Black-and-white photographic film is made of four basic layers plus two important adhesive layers, Figure 24-1. All of these layers added together measure about 0.005 inch (0.127 mm) in thickness. It is critical that the material used for each of the six layers meet consistent quality standards.

EMULSION. This is the light-sensitive layer of the film. It is made of minute (very small) particles of silver halides and gelatin. The silver halides are mixed (suspended) with gelatin, which acts as the vehicle. The mixture, now called the emulsion, is evenly coated on base material, Figure 24-2.

Silver halides are combinations of silver particles mixed with iodine, chlorine, or bromine. The very small crystals formed by combining silver with

bromine are the most sensitive. This makes them the most popular for photographic use. Silver bromine crystals measure 500 to 4000 nanometers (0.0005 to 0.004 mm) across. They are distributed throughout the gelatin vehicle as evenly as possible.

BASE. This is the flexible support upon which the other film layers are coated. Two materials are used for the film base of most films. These are *cellulose triacetate* made from wood pulp and *polyethylene terephthalate* made from ethylene glycol and other organic chemicals. Both of these base materials are used throughout the world for film manufacture. The major requirements of film base

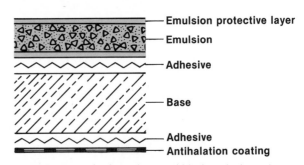

FIGURE 24-1. The basic layers of black-and-white photographic film.

material are that it be optically transparent, strong, and unaffected by water and processing chemicals.

PROTECTIVE LAYER. This is a thin layer of gelatin that serves to protect the light-sensitive emulsion. Flexible film is subjected to considerable handling and rubbing while in the camera and during processing. The gelatin layer does an excellent job in keeping the emulsion from being damaged.

Gelatin is very useful for many photographic materials. It is made from cattle hides, hooves, and bones. When hot, it is a liquid and can be coated onto a surface such as a film emulsion. After the hot gelatin cools and dries, it becomes hard, smooth, and transparent. This feature makes it excellent for needed light transmission for film.

ANTIHALATION LAYER. Sometimes underestimated in value, this layer absorbs light that passes through the emulsion. If this did not occur, the light would reflect back to the emulsion and give it a second, unwanted exposure. This layer is essentially a thin coating of dyed gelatin that also serves as an anti-curl material. *Halation* refers to the spreading of light beyond its normal boundaries. Thus, anti or keeping halation from happening.

ADHESIVES. Two thin layers of special glue-like gelatin that are coated on both sides of the base material. Sometimes these layers are called "subbing" because they form a foundation for the emulsion and antihalation layers. It is important that these two layers be completely transparent.

FILM CHARACTERISTICS

Different films are categorized according to six important characteristics. Each characteristic is carefully planned and measured during manufacture.

SENSITIVITY TO LIGHT. Film is rated according to its reaction to light, Figure 24-3. This is

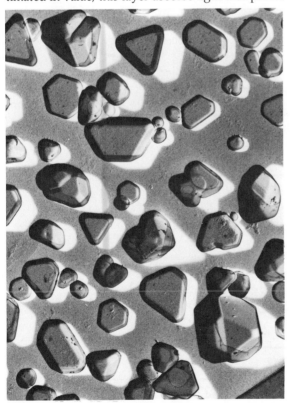

FIGURE 24-2. A pictorial representation of minute silver halide crystals magnified 10,000 times. *Ilford, Inc.*

ISO	DIN	Slow	Medium	Fast	Ultra-fast
1000	32				▓
900	31				▓
800	30			▓	
650	29			▓	
500	28			▓	
400	27			▓	
320	26		▓		
250	25		▓		
200	24		▓		
160	23		▓		
125	22		▓		
100	21		▓		
80	20	▓			
64	19	▓			
50	18	▓			
40	17	▓			
32	16	▓			
25	15	▓			

FIGURE 24-3. Black-and-white film speeds categorized according to their sensitivity to light.

normally referred to as film speed, and each film is identified by ISO and DIN numbers (see unit 15).

SENSITIVITY TO COLOR. Film emulsions are grouped according to their sensitivity to different wavelengths of light, Figure 24-4. *Blue-sensitive films* are sensitive to blue and limited ultraviolet wavelengths. These emulsions are usually very slow and have limited use in regular photography. *Orthochromatic* emulsions are sensitive to near ultraviolet, blue, and green light. Film in this group has limited but valuable use in regular photography.

Panchromatic film is sensitive to near ultraviolet and all three primary colors of light—blue, green, and red. A *spectrogram* or line diagram of part of the electromagnetic spectrum helps to make this clear, Figure 24-5. This film emulsion must be handled in total darkness. Almost all of the film

used in continuous-tone photography is of this type because subjects captured on film are made of every color imaginable. *Infrared film* is sensitive to all visible light waves, including those before 400 nm and beyond 700 nm. This film produces interesting results and is useful in research and scientific work. It is necessary to use a red filter on the camera lens when taking pictures with infrared film to block out blue and green light waves. If this is not done, the picture will look as if it was taken with panchromatic film.

IMAGE CONTRAST. This characteristic refers to the ability of a film to record the various degrees of gray that make up everything visible. Continuous-tone films are designed to record images from very light gray to very dark gray, Figure 24-6. Some films do a better job than others. This makes it important

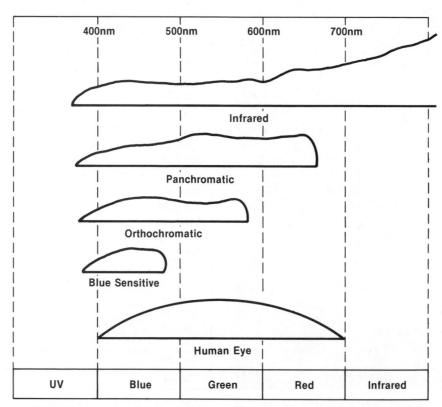

FIGURE 24-4. Color sensitivity of photographic film compared to the human eye.

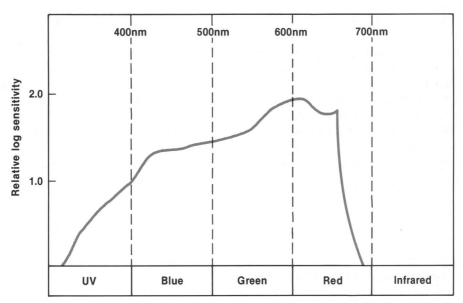

FIGURE 24-5. A test spectrogram of the color sensitivity of a panchromatic black-and-white film.

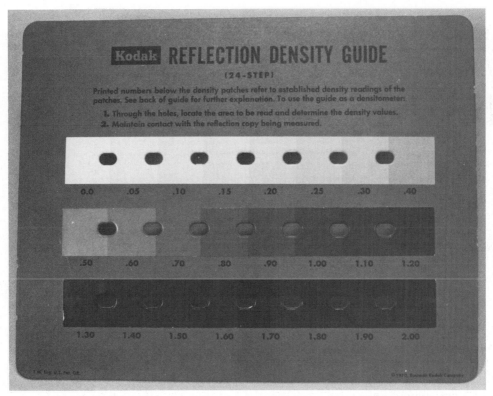

FIGURE 24-6. A gray scale illustrates the various tones from very light gray to very dark gray that can be recorded with continuous-tone film. *Eastman Kodak Company*

for a photographer to study the technical data prepared for the film and to conduct tests under controlled conditions.

Films that record a limited number of gray tones are called *high contrast*. This film type records only the medium-dark to dark gray tones. There is some use for this film in regular photography, but most of it is used within the graphic arts industry for printing purposes. See Figure 24-7 for a comparison of results with continuous-tone and high contrast films.

GRAIN OF EMULSION. *Grain* has a direct relationship to the speed of the film. A fast film contains larger silver halide crystals than a slower film. All film has grain because the clumping or grouping of the silver halides causes a visual pattern to evolve. Grain generally becomes a problem when an enlarged print (8″ × 10″ or larger) is made from a small negative of a fast film such as ISO 400. There are times when visible grain is desired in the finished photograph (see unit 15).

RESOLVING POWER. How a film emulsion records test images of very narrow, closely spaced lines is referred to as *resolving power*. Some films can record these lines and spaces better than others due to the emulsion mixture and other layers of a given film. Manufacturers test their films for this characteristic, but the user seldom has any need to do so.

CONTINUOUS-TONE FILM

HIGH-CONTRAST FILM

FIGURE 24-7. A comparison of the same scene photographed with continuous-tone film and with high contrast film.

IMAGE SHARPNESS. As with resolving power, this characteristic is tested by the manufacturer. The test measures the ability of a film to record a single "knife-edge" line. This characteristic is often referred to as the *acutance* of a film.

LATENT IMAGE

At the moment light strikes photographic film, an invisible or latent image is formed. This hidden image is created when the millions of photons (light) reach the film emulsion that is made of millions of minute silver halide crystals.

When a minimum of four photons strike a single silver halide crystal, a sequence of events take place. The events prepare the silver halide crystal to turn to black metallic silver when developing chemistry surrounds it, Figure 24-8. The silver particles (atoms) move together in clumps to form the visible image when developed. This is due to the sensitivity specks that are formed in an exposed crystal. The sensitivity specks are points where the silver particles have become trapped and have become activated by the photons. A *latent image* will remain in film for an indefinite period of time. Proper storage in cool, dry conditions permits latent images in some film to remain for months and even years before being made visible with developing chemistry.

CHARACTERISTIC CURVE

Describing how film reacts to varying amounts of light can be described with a graph called a *characteristic curve*, Figure 24-9. Developing time, temperature, and strength of the developer also can be charted on a characteristic curve. To provide meaning, only one variable should be tested at a time then plotted on the graph.

The characteristic curve in Figure 24-9 compares the density of developed film and the differing amounts of exposure. Continuous-tone film exposed to various amounts of light reacts differently than may be expected. Short exposures cause little density to occur until a given point. This is called the toe of the curve. The reverse is true when long exposures are given to film. Here the increase in density is negligible. This area of the curve is referred to as the shoulder.

The area of the curve that increases consistently in a straight line is simply called the straight-line portion. This is where the different exposures make a difference in the film's density. This, then, becomes the parameters of acceptable exposure.

This type of graph is sometimes called the H & D Curve. It was so named for the people, F. Hurter and V. C. Driffield, who began conducting research in this area of film exposure and development during the 1890s. Today, it is most often referred to by its generic name, characteristic curve.

FIGURE 24-8. Magnified 20,000 times, silver halide crystals are in the process of being made visible after 16 seconds in developing chemistry. *Ilford, Inc.*

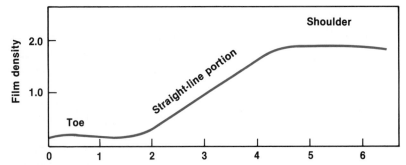

FIGURE 24-9. A characteristic curve is a graph that can be used to show the relationship between film density and the amount of exposure.

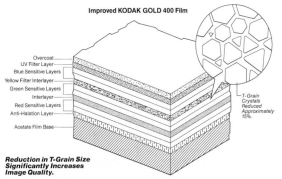

FIGURE 24-10. After considerable testing, researchers determined that reducing the size of silver halide crystals improved the imaging characteristics of this color film. *Eastman Kodak Company*

IMPROVING FILM AND PACKAGING

Researchers employed by photographic film manufacturers are constantly making improvements in silver-based films. Three such improvement areas include image sharpness, exposure latitude, and preprocess stability. Image sharpness or quality has been improved by reducing the grain size in color negative film, Figure 24-10. This result was accomplished only after much research and testing of the various emulsion mixtures and silver-halide grain crystal sizes.

Exposure latitude means that the amount of image light reaching the film can vary a fairly large amount and an acceptable exposure still will be obtained. Preprocess stability refers to the consistency

of the film emulsion from the time it is manufactured and when it is used by the consumer. For quality results, the emulsion should remain stable until it is chemically processed.

The consumers of both black-and-white and color films are kept informed of these improvements through advertising literature and package labeling. Film users, whether they be amateur or professional photographers, should constantly watch for new photographic products to come available.

A valuable new addition to film packaging is the DX CODE, Figure 24-11. This system is useful for both the photographer and the processing laboratory. The *bar code* contains processing information for the automatic detection of manufacturer, process, film type, film speed, and film length. The *checker code* is useful with cameras having built-in electric contacts that are designed to automatically detect the film speed, film length and the exposure range.

Also, on suitably equipped cameras, a view window permits the user to see the film type, speed, and number of frames. This feature is critical for photographers who change films often because it is always important to know what film is loaded in the camera.

HISTORICAL HIGHLIGHTS

Flexible photographic film first became available for purchase by the public during the 1880s. Prior to that time, light-sensitive emulsions were coated on

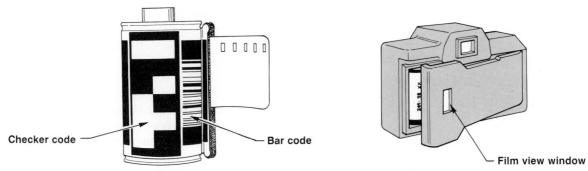

FIGURE 24-11. There are two significant components of the DX code; the *Bar* code is utilized by the processing laboratory, the *Checker* code is utilized by specially equipped DX code cameras, and the film view window is for the photographer. *AGFA Corporation*

rigid glass plates. This made photography a "strong man's" activity due to the weight of the light-sensitive glass film plates. An 1887 newspaper ad-vertisement depicts the promotion of photographic materials and services, Figure 24-12. Note the company name and one of its main products—glass dry plates.

SUMMARY

Continuous-tone film is a scientific wonder. Science and technology have been combined to create and manufacture a product that permits the recording of both visible and nonvisible images on film. The ease with which images can be recorded today makes for little comparison with films prior to 1900. For a thorough knowledge of a particular film, technical data sheets are available from the manufacturer.

REVIEW QUESTIONS

Answer these questions to test your knowledge of the unit content.

1. Name the six layers of black-and-white continuous-tone film.
2. Which film emulsion sensitizing element is the most sensitive?

   A. Iodine   C. Chlorine
   B. Bromine   D. Fluorine

3. T/F Cellulose triacetate is an excellent material for the film base layer because of its strength and transparent characteristics.

FIGURE 24-12. By 1887, the fledgling business offered a variety of products and services, as this advertisement indicates. Glass dry plates, papers, and films were among the products offered to the public. *Eastman Kodak Company*

4. What material is common to five of the six layers of black-and-white continuous-tone film?

5. Which type of film is not sensitive to the red wavelengths of light?

    A. High contrast   C. Continuous-tone
    B. Panchromatic   D. Orthochromatic

6. T/F The film characteristics of image contrast and acutance mean about the same thing.

7. Briefly describe the condition known as "latent image."

8. Name the three parts of the characteristic curve.

9. T/F The "bar code" information on a roll of film with the DX code helps the photographer make better exposures.

10. T/F Flexible photographic film is a product of the 19th century.

———— UNIT 25 ————

SAFETY IN THE DARKROOM

OBJECTIVES   *Upon completion of this unit, you will be able to:*
- *Select and wear proper safety equipment when working with photographic chemistry.*
- *Analyze a darkroom and determine if it is a safe working environment.*
- *Safely mix and store photographic chemistry.*
- *Describe (demonstrate if necessary) chemical splash eye wash-out procedure.*

KEY TERMS   *The following new terms will be defined in this unit:*

Chemical Concentrate   Eye Wash Station
Contact Dermatitis   Working Solution

INTRODUCTION

Working in a darkroom is very enjoyable and educational. The rewards of processing a roll of film or making a set of enlarged photographs are many. As in any working environment, there are dangers that wait at every corner unless precautions are taken. Photographic darkrooms are no exception. Anyone working in or visiting a darkroom must follow standard safety practices.

PERSONAL PROTECTION

The human body is vulnerable to injury from many sources. Punctures, bruises, burns, irritations, and sudden shocks can occur in a darkroom. It is always important to follow safe working practices while in and around any photographic darkroom environment.

1. Develop and practice an attitude of safety. Accidents can and do happen to people from all walks of life.

2. Wear eye protection when mixing and working with chemical concentrates, Figure 25-1. Be certain the goggles have been sanitized and thoroughly cleaned before putting them on.

3. Wear rubber-coated gloves when mixing working solutions from concentrates. *Working solutions* are those that have been diluted with water and are ready or near ready to be used in processing film and paper. Disposable medical gloves are excellent. After each use, rinse the gloves in water before taking them off. Once off, thoroughly wash the gloves inside and out with a mild acidic soap. Hang them to dry before using again.

4. Use print tongs to transfer photographic prints from one processing tray to another, Figure 25-2. This keeps the processing chemistry from ever touching gloved or ungloved hands.

5. Become very familiar with the layout of the darkroom with the white lights on. Under safelight conditions, it then will be possible to move about the darkroom without causing bumps and bruises to your body.

6. Keep cabinet doors and drawers closed except, of course, when something is being removed or replaced. Serious head and body injuries can easily occur by hitting open doors and drawers.

7. Protect your respiratory system by wearing a mouth and nose protective mask when mixing working solutions from liquid and powder concentrates. Dust from powder chemicals and fumes from liquid chemistry can be very harmful to the lungs. Protective masks are available from most paint and wall covering retail stores.

8. Always wear a rubber-coated apron when working with or around chemical solutions. Splashes and spills can occur at any time even under the best of conditions. Natural or synthetic fibers used in clothing absorb liquid solutions quickly. This causes the chemical solution to not only contact the skin but to remain in contact. Remove saturated clothing as quickly as possible and wash the contaminated area with a mild acidic soap and water.

9. Avoid *contact dermatitis* by following acceptable darkroom safety practices. Skin that becomes red, swollen, or begins to itch may be an indication

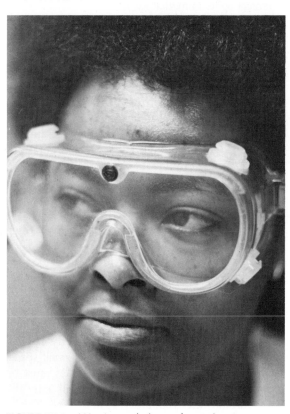

FIGURE 25-1. Wearing splash-proof goggles is a must when mixing or working with photographic chemical concentrates.

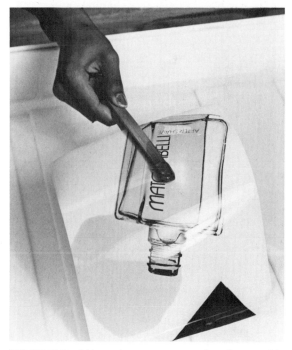

FIGURE 25-2. Using print tongs to transfer photographic prints from one tray to another keeps the chemistry from irritating the skin.

that you have the condition identified as contact dermatitis. Another symptom includes tiny white water blisters where the chemistry has contacted the skin. Do not treat these symptoms yourself. See a physician for an accurate diagnosis and proper treatment.

10. Be a well-dressed photographic darkroom worker, Figure 25-3. Wear clothing that completely covers your body. Always wear complete shoes and stockings. Sandals and thongs are taboo in a darkroom. Aprons, goggles, rubber gloves, and a respiratory mask are always in fashion in a darkroom, whether it be large or small.

DARKROOM CONDITIONS

The layout of a darkroom is important to safety. People who plan new or remodeled darkrooms

FIGURE 25-3. A well-dressed darkroom worker is well-protected from chemical hazards.

must be aware of the many factors that help make the darkroom environment safe.

1. Know the layout of the darkroom. Panchromatic film must be handled in total darkness; thus, know where cabinets, equipment, chemistry, and tools are located. Being unfamiliar with a darkroom can lead to an injury.

2. Maintain the brightest lighting possible so it will be possible to see and work conveniently. It will be necessary to determine if the safelights do not expose photographic paper.

3. Floors must be free from small or loose objects. Make certain the waste basket is positioned in or under a cabinet or table. Keep the floor dry because a wet floor is slippery. Wipe up water and chemical spills immediately.

4. Doors and walls should be painted or covered with a light-colored material. When it is possible to use safelights for making prints, the available "safe light" will be reflected throughout the darkroom. Black doors and walls absorb the available light making it difficult to see.

5. Position door handles and locks in logical locations. Also, make certain door handles are free of sharp edges and rough surfaces. Skin abrasions and cuts can easily occur when reaching for an unsafe door handle.

6. Locate equipment where it is easily accessible. If it is necessary to reach a long distance to activate a machine such as a film or print drier, then it may be unsafe.

7. Electrical wiring and switches must be properly grounded. Loose wires and worn switches have no place in a photographic darkroom. **NEVER** touch an electrical wire or switch with wet hands or feet. Water conducts electrical current extremely well, and an electrical charge into the body can kill.

8. Ventilation is important in any small space, especially in a darkroom. Chemical fumes must be removed from the work area or they will enter the body through the respiratory system. The air in a darkroom should be completely changed every 8 to 10 minutes to be safe.

MIXING CHEMISTRY

Anyone responsible for mixing photographic chemistry must know and follow safe procedures. *Chemical concentrates* contain acids and alkalis of such strong proportions that adequate protection must be made.

1. Ask permission of your instructor or supervisor before mixing working chemistry from a liquid or powder concentrate.

2. Wear the appropriate safety apparel—goggles, gloves, mask, and apron—when mixing or working with photographic chemistry.

3. Mix chemistry in open stainless steel containers. This should be done for two important reasons. First, if dropped or knocked over, the steel container will not break. Second, chemical contamination will not occur assuming the container has been properly cleaned prior to use. Some plastic containers may cause contamination because the chemical concentrate has remained in the pores of the plastic.

4. Read all instructions completely and carefully, Figure 25-4. Information is available on how to mix the concentrate to a working solution. More importantly, warnings are provided by the manufacturer regarding the dangers of working with the specific solution or powder. **NEVER** drink or allow anyone to drink photographic chemistry. This is especially true if small children have access to a darkroom.

5. Always pour an acid base solution into water. Also, pour slowly and with constant stirring. Pouring water into an acid can cause the solution to boil. Your hands and face could be seriously burned if the acid and water mixture should spatter.

6. Always pour and mix photographic chemicals below eye level. This greatly reduces the chance of their splashing on exposed skin and eyes.

7. Know the location and how to use an *eye wash station*, Figure 25-5. Read the instructions and request a demonstration of the system before working with photographic chemistry. A small wall mounted stainless steel sink containing two upward flowing water spouts also works well.

8. Eyes should be washed immediately with solution from the eye wash station or with cool running water if chemical splashes to the eyes should occur. Pull back the eyelids and allow the running water to flush away the chemicals for a minimum of 15 minutes. This will be uncomfortable and even painful, but it is the only way to save eyes from permanent chemical splash burns.

FIGURE 25-4. Information and safety instructions should always be read before working with photographic chemistry.

FIGURE 25-5. An eye wash station should be located within 25 feet (7.6 m) of any photographic chemical work area. *Bel-Art Products*

FIGURE 25-6. Photographic chemistry should always be stored on low-level shelving.

STORAGE OF CHEMISTRY

Proper storage of concentrate and working solution chemistry is very important to photographic safety. Anytime there are one or more persons using chemical solutions, it is critical that attention is given to where and how they are stored.

1. Plastic bottles, large or small, should always be used to store photographic chemicals. Glass containers of any type or size should never be found in a darkroom. It is very easy to drop and break glass containers because they become slippery when wet. Be certain to replace and secure the plastic cap on each bottle after use.

2. Store all chemical containers, large and small, on low shelving, Figure 25-6. A good guideline to follow is to store chemicals on shelves below eye level.

3. Label all chemical containers with clear, concise information. Never leave a chemical mixture in a sealed container without adding an identification label. Cover the label with clear sealing tape to keep it dry and securely fastened to the container.

SUMMARY

Darkroom safety is a matter of forming and following good work habits. Follow the guidelines given in this unit so that your photography experiences will be pleasant and safe. Remember—safety is an attitude. People who believe personal safety is important will also practice it.

REVIEW QUESTIONS

Answer these questions to test your knowledge of the unit content.

1. Why is an attitude of safety important when working in a darkroom?
2. Rubber-coated _____ should be worn when mixing photographic working solutions from concentrates.
3. T/F It is appropriate (safe) to transfer photographic prints from one chemical tray to another with your fingers.
4. What is the skin medical problem that can sometimes bother photographic darkroom workers?
5. T/F Anyone working in a darkroom should be very familiar with its complete layout.
6. Darkroom ventilation air should be completely changed every _____ to _____ minutes.

   A. 4 to 6   C. 12 to 15
   B. 15 to 20   D. 8 to 10

7. It is best to mix photographic chemistry in _____ containers.

   A. Stainless steel   C. Plastic
   B. Glass   D. Porcelain

8. T/F Acids should always be poured into water.
9. How long should an eye be flushed with a special wash solution or water after receiving a chemical splash?

   A. 5 minutes   C. 15 minutes
   B. 10 minutes   D. 20 minutes

10. T/F It is appropriate to store plastic bottles of photographic chemistry at or above eye level.

CHEMISTRY FOR FILM PROCESSING

OBJECTIVES   *Upon completion of this unit, you will be able to:*
- *Explain the purpose and contents of film developers.*
- *Identify the purpose and contents of stop bath solutions.*
- *Name the purposes and contents of fixers.*
- *Describe the two main film processing finishers.*

KEY TERMS   *The following new terms will be defined in this unit:*

Accelerator	Indicator Stop Bath	Preservative
Chemical Fog	Metol	Reducing Agent
Developer	MQ Developer	Replenishing Developer
Film Finishers	Phenidone	Restrainer
Fixer	PQ Developer	Stop Bath
Hydroquinone		

INTRODUCTION

There are two primary purposes of film processing chemistry: to make the latent image visible and to make the visible image permanent. Specific chemical agents are used to increase or decrease the image contrast, while other chemicals are used for image enhancement in other ways. Film processing chemistry can be broadly grouped into four categories: developer, stop bath, fixer, and finishers.

DEVELOPERS

The purpose of a *developer* is to make the latent image in film emulsion visible. It does this by turning the exposed silver halide crystals to black metallic silver. The developer must have the capability to distinguish between the exposed and non-exposed silver halides.

Photographic film developing chemistry contains four basic parts: reducing agents, accelerators, preservatives, and restrainers, Figure 26-1.

REDUCING AGENTS.   The *reducing agent* is also called the developing agent. This portion of a developer is that which converts the exposed silver halides to metallic silver. Over 400 chemical

compounds have been considered for use as photographic reducing agents. Only a few of this large number are of practical value. The three compounds

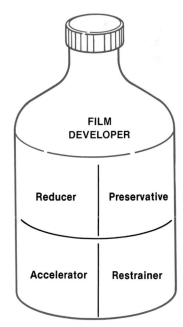

FIGURE 26-1.   The four basic ingredients (agents) of photographic film developer.

most often used in black-and-white film developers are Metol, Phenidone, and hydroquinone.

Metol is also known by other company names as Elon, Genol, Pictol, Photol, Planetol, Rhodol, Scalol, and others. It is the most rapidly acting of the three reducing agents. Used alone, it reduces or develops some of the silver halides that were only slightly exposed to light. This causes a condition known as *chemical fog.* This means that the lightest areas on the film negative (darkest on the print) will be developed too much, Figure 26-2.

Phenidone was discovered by Ilford Ltd. in 1940. If used alone, it too produces chemical fog because of its fast-acting capabilities. *Hydroquinone* is a slow acting reducing agent and is often used alone in developers for high contrast film. It is capable of producing very high contrast in the film emulsion.

A combination of reducing agents is used in most continuous-tone, black-and-white film developers. In this way, the good features of each agent can be emphasized. The initials *MQ* are used for Metol and hydroquinone developers and *PQ* for Phenidone and hydroquinone developers. Many other combinations also are used.

ACCELERATORS. The accelerator component of a developer is sometimes referred to as the activator. Without an accelerator, the reducing agents are rather inactive. To make them active, an alkaline chemical is used.

PRESERVATIVE. This ingredient within the developer solution keeps the reducing agents and accelerator from spoiling. Sodium sulfite is a commonly used preservative.

RESTRAINER. Potassium bromide or benzotriazole is included with the reducing agent, accelerator, and preservative to keep the developer from causing chemical fog. It serves to restrain the reducing agent from attacking the unexposed silver halides. In effect, it serves to protect the unexposed silver halides.

CATEGORIES AND TYPES OF DEVELOPERS

Developers can be divided into three general categories: general purpose, fine grain, and high energy. The general purpose developers are just that. They can be used for many different continuous-tone films and give good results. The fine grain developers produce small, sharp-edged silver halides. This helps to give extremely sharp enlargements from small film sizes like 35-mm.

High-energy developers are strong and fast acting. They are formulated to bring out the weakest of images in a film emulsion. Underexposed film

FILM NEGATIVE

PHOTOGRAPHIC PRINT

FIGURE 26-2.   Chemical fog results when the developer reducing agent attacks the unexposed silver halides.

can be effectively saved with developers containing the right combination of reducing agents.

There are several types of film developers. Most developers for hand processing are of the one-time use type. This means that the right mixture and dilution is made and used to develop a single roll or several rolls of film at one time. After development is completed, the developer is discarded down the drain.

Replenishing developers are used most often with machine film processing. After the developer is used, a replenisher is added to bring it back to full strength. The replenisher strengthens the reducing agent within the developer. Great care must be exercised when adding replenishers so that the correct developer strength is obtained.

Developers are available in liquid or powder concentrates. The type used varies with the end use and the wishes of the darkroom worker. There are many types and brands of film developers, Figure 26-3. This gives the photographer a variety of choices. It is important to begin film developing experiences with a basic, general purpose developer. Become well acquainted with it before changing and experimenting with other developers. Darkroom chemistry kits to individual bottles of chemistry concentrates are available, Figure 26-4.

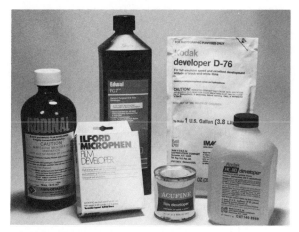

FIGURE 26-3. Many types and kinds of continuous-tone developers are available for the photographer.

STOP BATH

The purpose of stop bath solution is to neutralize the action of the developer solution. *Stop bath* is a mild mixture of acetic acid that counters the developing activity of the alkaline accelerator in the developer solution. It is possible to mix stop bath solution from caustic glacial acetic acid or from 28% acetic acid.

It is strongly recommended that a commercial stop bath solution be used. Most of these stop baths have a feature known as an *indicator*. This is a coloring agent of yellow dye that turns purple when the stop bath solution is exhausted. This feature is not usually needed for roll film processing because a quantity of stop bath should only be used one time then properly discarded. The indicator feature does serve good use when processing sheet film and photographic prints in a tray. Commercial stop bath is very economical to purchase in a concentrate and has a very long shelf life.

Water is used by some photographers as a stop bath. This is a poor practice because water

FIGURE 26-4. Darkroom processing chemistry is available in several container sizes of chemical concentrates. *Heico Chemicals, Inc.*

provides no chemical neutralizing action as does the acid-based stop bath. Water only washes away the developer, thus it is much slower and less reliable.

FIXER

Fixer solution has two main purposes. First, it serves to dissolve the silver halides that were untouched by the reducing agents of the developer. Second, the hardening agent strengthens the silver halides that were turned to black metallic silver.

After photographic film and paper have been developed and even subjected to stop bath, the unexposed and undeveloped silver halides still remain sensitive to light. The fixer solution must be used to remove them, otherwise the remaining developer solution and light will turn the entire film black.

Sodium thiosulfate is the major ingredient in most fixing baths. It has the capacity to dissolve

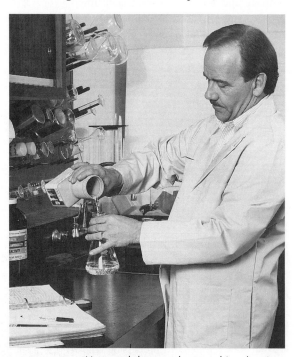

FIGURE 26-5. New and better photographic chemistry is being pursued constantly. *René C. Gallet*

unexposed and undeveloped silver halide crystals. Ammonium thiosulfate is used in some "rapid fixing baths," because of its ability to work more rapidly. The hardening agent most often used is potassium alum. A small proportion of acetic acid also is included. This does enforce the neutralizing action of the stop bath solution.

FILM FINISHERS

Photographic chemistry in this category includes the film washing and drying solutions. After the roll or sheets of film have been developed, stopped, and fixed, it is important to remove all of the acid and alkaline base solutions. Water is the primary substance used, but to save water and washing time, a washing aid is used. This solution is called a hypo clearing agent. It quickly dissolves remaining unexposed and undeveloped silver halides, and breaks up and removes any of the fixing bath.

Drying aids are also known as wetting agents. The purposes of using a drying agent on film are to decrease water surface tension, minimize water marks and drying streaks, and speed drying. Water streaks and spots can be a serious problem on a developed roll or sheet of film if a drying aid is not used.

SUMMARY

Chemistry for black-and-white film processing has taken years to formulate and perfect. Chemists are constantly searching for new and better compounds to use in film processing, Figure 26-5. The photographic darkroom worker must take time to become acquainted with the many brands and types of film processing chemistry. In this way, quality, efficiency, and economy can be achieved.

REVIEW QUESTIONS

Answer these questions to test your knowledge of the unit content.

1. What are the two primary purposes of film processing chemistry?
2. Name the four basic parts of a photographic developer.
3. Which of a developer's four basic parts turns the latent image to black metallic silver?
4. Of the common developing agents, which acts most rapidly?

   A. Phenidone   C. Metol
   B. PQ   D. Hydroquinone

5. T/F The accelerator or activator agent in a developer solution has an alkaline base.

6. Chemical fog is reduced or eliminated by which developer agent?
7. T/F Replenishing developers are those that restrengthen themselves in a container after being used.
8. T/F Indicator stop bath contains a special red dye that can turn purple.
9. Why is it necessary to use a fixer solution when processing photographic film?
10. T/F Film washing and drying solutions help save time and processing cost.

— UNIT 27 —

FACILITIES FOR FILM PROCESSING

OBJECTIVES *Upon completion of this unit, you will be able to:*
- *Specify equipment needed for quality film processing.*
- *Assemble the various tools useful in processing roll or sheet film.*
- *Design a film processing darkroom.*

KEY TERMS *The following new terms will be defined in this unit:*

Film Clip	*Graduate*	*Processing Tank*
Film Developing	*Loading Room*	*Squeegee*
Film Processing	*Processing Room*	

INTRODUCTION

Good facilities are important for efficient and consistent *film processing*. Basic equipment, tools, and a room which can be made 100% dark are essential to the task of making latent images visible in film. Also, every possible effort to keep the facilities neat and clean must be made by everyone involved in film processing.

EQUIPMENT

Ten pieces of equipment are needed to process black-and-white film. Various quality levels are available for each. It is wise to obtain the medium to high-quality equipment whenever possible. This always helps to ensure success in processing roll and sheet film.

SINK. A sink should be large enough to be conveniently used by at least one person, Figure 27-1. The features should include a deep sink to help keep from splattering chemistry and water, and a work area drain board. Also, two water faucets are very useful. When one faucet is being used for washing a developed roll of film, another faucet is available. Storage space under the sink is essential for both chemistry, equipment, and tools.

FIGURE 27-1. A well-designed and equipped sink suitable for both film and print processing. *Kreonite, Inc.*

TEMPERATURE CONTROL. A good quality film processing sink will always contain a temperature control water system. This is true with the sink shown in Figure 27-1. A hot–cold water mixing value that can be regulated with a control knob and thermometer helps to maintain quality film processing. Water and chemical temperature are critical to consistent results.

TIMER. An accurate timer that can be set for both minutes and seconds is needed, Figure 27-2. The numbers on the timer face should be large and easy to read. Be certain the timer contains a three-wire grounded electrical plug and convenient switch locations.

FIGURE 27-2. A timer is a necessary piece of darkroom equipment. *Dimco Gray Company*

CLOCK. A large-faced wall clock is very useful for timing of selected processing stages. Make certain it includes a large, easy-to-see second hand.

PROCESSING TANKS. A *processing tank* is the most important piece of equipment for film processing. It must be light-tight and be designed for rapid entry and exit of the processing chemistry. There are three categories or styles of tanks, Figure 27-3.

STAINLESS STEEL TANK   APRON PLASTIC TANK   RATCHET PLASTIC TANK

FIGURE 27-3. The three standard styles of roll film processing tanks.

FIGURE 27-4. A ratchet reel that can be easily loaded with a roll of film. *Jobo Fototechnic, Inc.*

The long-standing stainless steel unit includes the tank, reel, lid, and cap. A second tank style includes a plastic apron with ruffled edges to wrap the film around. The tank is made of hard plastic, except for the steel plates used to hold the apron and film in position inside the tank. A third and popular tank style includes a ratchet reel, Figure 27-4. One of the two side plates can be moved back and forth to load the film. As with all film tanks, it is important that the reel or apron keep the film from touching.

Sheet film tanks, film hangers, and film loaders have convenience features, too, Figure 27-5. A darkroom worker should try different tanks and select one that can be used with consistent results. A feature that is sometimes overlooked is the ability to wash the film. The best type is one that permits a hose to be placed into the center of the tank, Figure 27-6. The running water is forced to

FIGURE 27-5. Sheet film can easily be loaded into a reel with a special loading device. *Jobo Fototechnic, Inc.*

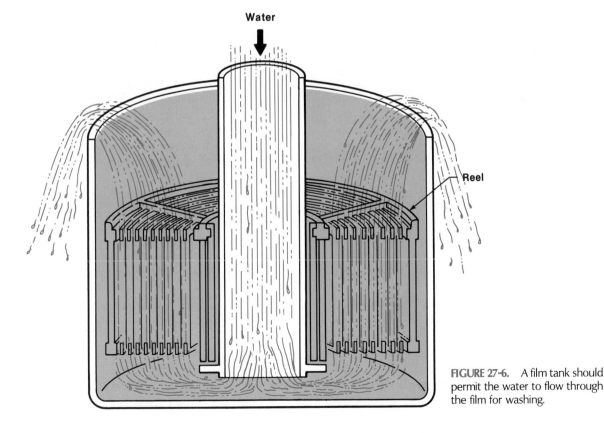

Water

Reel

FIGURE 27-6. A film tank should permit the water to flow through the film for washing.

FIGURE 27-7. Compressible, brown plastic chemistry containers help keep developer and other photographic chemistry from oxidizing. *Falcon Safety Products Inc.*

the tank bottom, up through the reel with film, and out over the top edge. This permits the water to remove the used chemistry from the film and carry it quickly out of the tank.

Roll film tanks are available for single and multiple roll processing. The type and size used depends on the habits and needs of the darkroom worker.

MIXING BUCKET. A stainless steel mixing bucket is useful for mixing working solutions from chemical concentrates. A $1\frac{1}{2}$ to 2 gallon (5.7–7.6 liters) bucket is a suitable size for most situations.

GRADUATES. A minimum of three sizes of *graduates* are needed for hand processing roll and sheet film. The 8 oz (250 ml), 16 oz (500 ml), and 32 oz (1 liter) sizes are most useful. Be certain the graduates are made of plastic.

CHEMISTRY CONTAINERS. The best types of chemistry containers are made of dark brown plastic, Figure 27-7. The dark color is necessary to keep light from weakening the strength of the chemistry. A plastic container is also unbreakable; thus, it is safe for handling with wet hands. A compressible container helps keep air from making con-

tact with the developer and other photographic chemistry. Air causes oxidation, which means that oxygen combines with the chemicals and reduces their strength. To avoid contamination, always use containers for the same chemical. This is important even with careful cleaning of the containers.

FILM WASHER. Water circulation in and around film that has been developed and fixed is critical to complete washing. Several types of film washers are available. A popular type is shown in Figure 27-8. It accepts several reels of film at one time and can easily be attached to a water faucet.

FILM DRYER. A film cabinet dryer that circulates warm, filtered air around the wet film is the best type, Figure 27-9. Long film strips of 35-mm, 36 exposure and sheets of film can be hung inside the cabinet. It is important to be able to control the drying environment for film. Dust and dirt are enemies of photographic film.

TOOLS

Several tools are useful for roll and sheet film processing. Some of the tools can be obtained at a retail hardware store. Others must be purchased from a photographic equipment supply company.

SHEARS (SCISSORS). Almost any household type scissors will work fine. Make certain the blades are kept sharp and clean. A scissors must cut the film smooth and straight.

THERMOMETER. Several types of thermometers are available for photographic use, Figure 27-10. It is essential that a thermometer be accurate, easy to read, and have both Fahrenheit and Celsius scales.

FUNNEL. One or more funnels made of plastic or stainless steel is essential. Chemistry is used, mixed, and returned to containers on a constant basis in a film processing darkroom.

PROTECTIVE GLOVES. Several pairs of thin, rubber gloves are needed in any darkroom. A pair

should be worn when mixing working solutions from chemical concentrates.

CAN/BOTTLE OPENER. A beverage container opener is useful in removing the end cap from a

FIGURE 27-9. A film drying cabinet that circulates warm, filtered air around the hanging film. *California Stainless Mfg.*

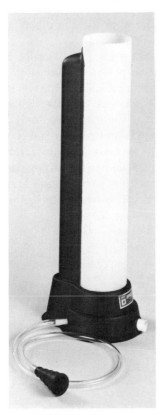

FIGURE 27-8. A film washer that can be used to wash several reels of film at one time. *Doran Enterprises, Inc.*

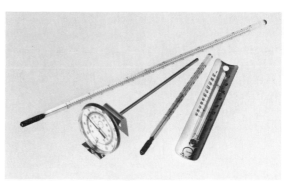

FIGURE 27-10. Several types of thermometers useful for photographic processing.

35-mm film magazine. It also is needed to cut V-shaped openings in the top of some chemical concentrate containers.

PLIERS. A standard, general purpose pliers is useful in any darkroom. Sometimes caps on chemical containers cannot be removed with the fingers. Also, a pliers is useful in opening some film cartridges.

FILM CLIPS. Either plastic or metal *film clips* are used to hang roll film for drying, Figure 27-11. Wood, clamp-type clothes pins can be used, but the wood is difficult to keep clean and free of contaminating chemistry.

MIXING ROD. Two or three mixing rods are useful when preparing chemistry for processing. Generally, they are made of plastic, $\frac{1}{4}''$ (6.4 mm)

in diameter, and 8'' to 12'' (20.3–30.5 cm) long. Some have small handles at one end.

SQUEEGEE. A *squeegee* used for roll film is designed to remove water from both sides at the same time, Figure 27-12. The multiple blades should be made of soft rubber so that the wet emulsion will not be scratched.

DARKROOM

The primary requisite of a darkroom is that it can be made 100% dark. Panchromatic film is sensitive to even the smallest amount of light that shines through a door opening. Doors into a darkroom must seal perfectly at all four sides. It is sometimes difficult to obtain a seal between the door bottom and the floor. Commercial sealing devices are available that can be fastened to the bottom of the door. When the door is closed, the rubber seal is forced down against the floor.

LOADING ROOM. A small room large enough for one person is useful for loading film into a developing tank. Once the film is in the tank, it can be brought out into the lighted darkroom processing area. The room need only be equipped with a working height counter and possibly a storage cabinet. The best location for the room is directly off the main film processing darkroom. This location permits film handling and loading in complete darkness and convenience to the main dark-

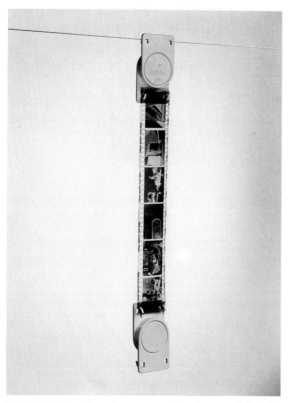

FIGURE 27-11. Film clips are used to hang roll film for drying. *Doran Enterprises, Inc.*

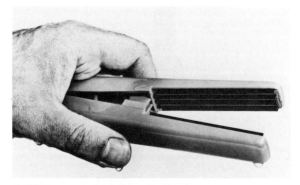

FIGURE 27-12. A squeegee used to remove water from roll film. *Jobo Fototechnic, Inc.*

room. Also, this area should be designated as a 100% dry work area. Water and photographic chemistry should never be allowed in the film loading room.

   PROCESSING ROOM.   This area is considered the main film processing darkroom. It must be large enough to hold a sink, film dryer, and counter work space. Technically, this room does not need to be made dark because *film developing* tanks keep all light from reaching the film. An exception to this nondark requirement is when sheet film is processed in an open tank or trays. Be certain to designate one area, such as a counter, as the dry area. In this way, information sheets and books can be placed there to provide processing data.

   TYPICAL DARKROOM LAYOUT.   Darkrooms should be designed and constructed in such a way that they can be used effectively, Figure 27-13. The size depends on the number of people who will use the darkroom at any one time. Most film

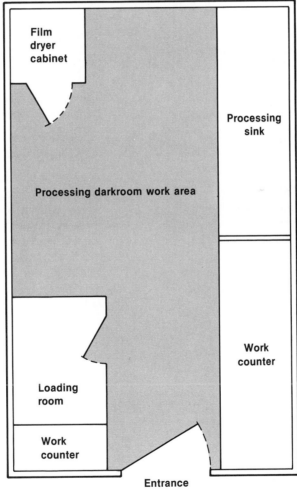

FIGURE 27-13.   Floor plan of a film processing darkroom in which three people can work effectively at one time.

processing darkrooms do not need to be very large. The time spent in a film processing darkroom is much shorter than that spent in a photographic print darkroom. This makes it possible for many people to cycle through in a short period of time.

HOME FACILITIES.   Darkrooms, both temporary and permanent, can be made in a home. The most common location for a temporary darkroom is a bathroom. Another is the kitchen or in a corner of the basement. The primary reason for these locations is the need for running water and a drain. Being able to make the room or area dark is also an essential ingredient..

Permanent home darkrooms should be located where environmental conditions, such as heat, light, dust, and dirt, can be fully controlled. Darkrooms can provide considerable enjoyment and opportunities for both young and old. Safety must be practiced at all times in both temporary and permanent home darkrooms.

SUMMARY

Equipment, tools, and darkroom facilities are necessary for photographic film processing. The quality, style, and brand of equipment and tools can vary a great deal. The main requirement is that the items and facilities identified in this unit be available. The darkroom worker has a major responsibility to maintain the equipment, tools, and facilities.

REVIEW QUESTIONS

Answer these questions to test your knowledge of the unit content.

1. Sinks for processing continuous-tone film work best when they are:

   A. Deep   C. Shallow
   B. Wide   D. Narrow

2. T/F A clock and a timer are essentially the same; therefore, there is need to have only one of these devices.

3. What is the most important piece of film processing equipment?

4. List the three types of roll film processing tanks.

5. A 16 oz graduate and a _____ ml graduate are about the same size.

   A. 250   C. 750
   B. 500   D. 1000

6. The best color for plastic photographic chemistry containers is:

   A. White   C. Red
   B. Yellow   D. Brown

7. A good quality thermometer suitable for photographic film processing will contain which two sets of scales?

8. Why should the blades of a squeegee used for roll film be made of soft rubber?

9. T/F A film processing darkroom must be able to be made 100% dark.

10. T/F Film processing darkrooms need to be rather large to accommodate the extensive equipment, tools, and personnel.

PREPARING FOR FILM PROCESSING

OBJECTIVES   *Upon completion of this unit, you will be able to:*
* *Prepare the darkroom for film processing.*
* *Load exposed film into a processing tank.*
* *Ready chemistry for film processing.*

KEY TERMS   *The following new terms will be defined in this unit:*
Film Leader
Spindle

INTRODUCTION

There are several steps that must be completed before chemistry and film meet. This preparation phase of black-and-white film processing is as critical as the actual chemical steps that are presented in unit 29. Remember to follow good working habits and safe practices.

PREPARING THE DARKROOM

Darkrooms, just like other rooms in a school, business, factory, or home, need to be cleaned often. Wipe counter tops, shelves, and cabinet doors with a cloth that contains a nonoil-based cleaner. Do the same to darkroom doors and walls. Use a vacuum cleaner to lift dust and dirt from the floor. If possible, place the exhaust of the vacuum cleaner outside of the darkroom. Some vacuum cleaners stir up dust on the floor. Also, shoes should be checked for cleanliness before entering the darkroom.

Film processing tanks, graduates, funnels, and other mixing tools must be kept clean. Leftover chemistry from a previous processing job can easily contaminate a new mixture. After each use, processing equipment and tools should be cleaned with plenty of running water. Also, it is valuable to wash all equipment and tools with a mild detergent soap and water. The same procedure as is done with the dinner dishes can be used. Be certain to rinse all soap off with warm water. Dry the equipment and

tools with a towel before putting them away. This procedure should be done on a weekly or monthly schedule.

Gather the clean and dry equipment and tools. Refer to unit 27 for the complete listing. The following equipment and tools are needed to load an exposed roll of film into a processing tank, Figure 28-1:
* Exposed roll of film
* Processing tank
* Bottle/can opener
* Scissors

Make a final check of the darkroom conditions. Turn out the lights and close the door to the loading room. Wait a few minutes, allowing the eyes to adjust to the darkness. Look closely at all sides and corners to determine if light is leaking in anywhere. If a light leak is found, it must be repaired before film loading can begin. If film loading will be done in the processing darkroom, this should be checked for light leaks too.

LOADING EXPOSED FILM

Loading exposed film into a processing tank is a delicate procedure. It must be done in total darkness; therefore, practicing the procedure in a lighted room is strongly recommended. Use a fogged strip of 35-mm film about the length of 20 to 36 exposures. Follow the procedural steps

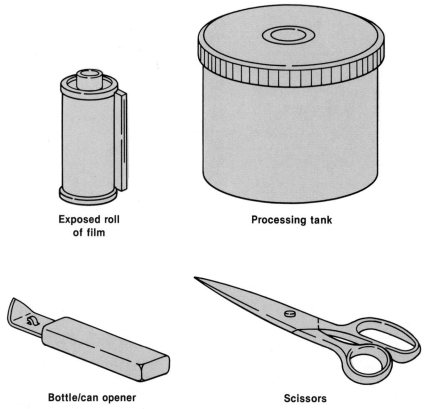

**Exposed roll
of film**

Processing tank

Bottle/can opener

Scissors

FIGURE 28-1. These four items are needed for loading exposed film into a processing tank.

several times until they become almost a natural habit. Going through the steps with your eyes closed is good practice, too. It is important to practice this entire procedure until all problem spots have been overcome.

1. Arrange the items to be used in "known" positions on the counter top in the loading darkroom, Figure 28-2. Take the processing tank apart and lay each part in a location where it can easily be found in the dark.

2. Make arrangements with another person to provide assistance in the event that a problem occurs during film loading. This step

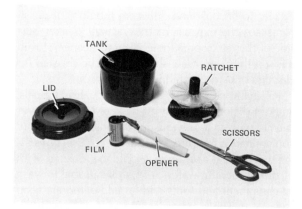

FIGURE 28-2. All items needed for loading exposed film into the processing tank should be arranged in a logical order.

FIGURE 28-3. Opening a film magazine with a bottle cap opener.

FIGURE 28-4. Pressing the spindle against a counter top to push the roll of film partly out of the magazine.

is important only until complete confidence is gained in the loading process.

3. Close the loading room door to establish 100% darkness.

4. Open the film magazine. Use the bottle cap opener to lift the end cap of the magazine, Figure 28-3. Be careful to not slip and injure fingers while hand holding the film magazine. On reusable magazines for bulk loading, the end cap can be pulled or twisted off without using the opener.

5. Remove the roll of film from the magazine. Press the *spindle* on the counter top to push the roll of film out of the magazine about 3/16" (4.8 mm), Figure 28-4. Take hold of the film roll with two fingers and thumb. Pull the film from the magazine making certain that it does not unroll. Hold the roll of film and spindle securely, but do not touch the image area. Film, exposed or unexposed, should always be handled by the edges. Oil from the skin will leave fingerprints on the film.

6. Cut off the *film leader*. Hold the film roll securely, and make the cut with a scissors, Figure 28-5. Be certain to cut the film straight across.

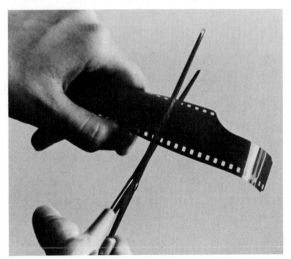

FIGURE 28-5. Cutting the leader from the exposed roll of film. This must be done carefully while in total darkness.

7. Trim the corners. For some styles of tank reels, it is important to trim the two corners at 45°, Figure 28-6. This must be done carefully so the corners are not cut back too

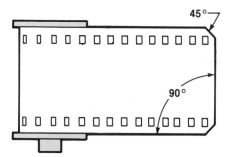

FIGURE 28-6. Trimmed corners make it easier to load film on many types of film tank reels.

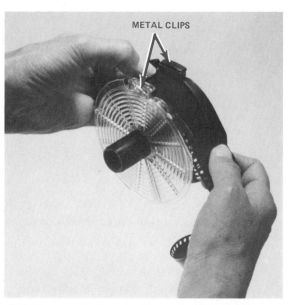

FIGURE 28-7. Pushing the film under the metal clips of the film processing tank ratchet.

far. Trimming these two corners makes the film easier to load.

8. Load film into the reel. Be certain the reel is clean and perfectly dry. The following procedure should be used to load a plastic ratchet reel.

a. Locate the groove opening lugs on both sides of the reel. Keep the index finger on this starting point. Hold the reel between the thumb and middle finger.

b. Push the trimmed end of the film into the groove beneath the opening lugs, Figure 28-7. Be certain the film is firmly caught below the metal clips. Remember to hold the film by the edges.

c. Release the spindle of film and allow it to hang. It must not touch the floor.

d. Ratchet the roll of film onto the reel, Figure 28-8. Move the two sides of the reel back and forth to advance the film. Stop when the spool nearly touches the reel.

e. Remove the spindle by cutting the film just ahead of the tape, Figure 28-9. Do not leave any of the tape on the film as it could cause contamination during processing.

f. Continue ratcheting the film onto the reel until the end is beyond the metal clips.

FIGURE 28-8. The film is advanced on the ratchet wheel by moving the sides back and forth.

FIGURE 28-9. Cutting the spindle and tape from the end of the exposed film that has been loaded into the processing tank ratchet.

9. Place the reel into the tank. Touch only the reel and not the film.

10. Attach the lid onto the tank. Firmly press the lid in position or screw it on depending on the kind of processing tank.

11. Turn on the lights or open the door to the loading room. There is no need to worry that the loaded film will be exposed to the light. The tank is light proof, assuming that the lid was properly attached. Place the loaded tank in a position where it will not be knocked off the counter.

12. Clean up the work area. Put away the tools, and discard the film clippings and one-time use magazine. If the magazine is of the reusable type, put it in its proper storage location.

PREPARING THE CHEMISTRY

Chemistry must be prepared with utmost care. Three chemical solutions must be mixed and placed in graduates before processing can begin. These are the developer, stop bath, and fixer. Eight ounces (250 ml) of solution are needed to cover

one roll of film in a standard roll tank. Double this amount if two rolls are being processed at the same time.

There are generally three levels of chemical strengths: concentrates, stock solutions, and working solutions. This is especially true with many of the developers. Companies formulate, manufacture, and distribute concentrates. They use this procedure to save on handling and shipping large amounts of weight. Also, some chemicals begin to break down rapidly when mixed with water.

The darkroom worker is responsible for preparing both the stock solutions and working solutions. Stock solutions are primarily associated with developers. They are prepared from the liquid or powder concentrates by adding a precise amount of distilled (mineral-free) water. From this point, additional water is added to make the working solution. This should be done just prior to use. Working solutions are mixed directly from concentrates with several photographic chemicals.

Select a developer for the film that is recommended by the manufacturer. Much of the time, three or more developers that can be used are listed in the film packaging, Figure 28-10. The temperature of the developer is very critical. A warm developer is more active than a cold solution. The optimum temperature for hand-processing chemistry is 68 °F (20 °C). As shown in Figure 28-10, the developing times are given based upon the standard developer solution temperature.

DEVELOPER. Most film developers require a further dilution from the stock solution. Follow the instructions provided by the manufacturer. For example, Kodak developer HC-110 (dilution B) requires 1 ounce (31.25 ml) of developer working solution to 7 ounces (218.75 ml) of water. This makes the needed amount of developer for one roll of film. Distilled water at 68 °F (20 °C) should be used. Tap water may be contaminated with mineral particles. This could cause serious developing problems.

Place the graduate on a flat surface within the "wet" area of the film processing darkroom. Pour in the correct amount of developer stock solu-

Developing Times in Minutes[†] at 68 °F (20 °C) Small (1-qt. or less) tank agitated at 30-sec intervals			
KODAK Developer	D-76 $5\frac{1}{2}$	MICRODOL-X 7	HC-110 (Dilution B) 5

[†] Unsatisfactory uniformity can result with development times shorter than 5 minutes.

FIGURE 28-10. Manufacturers list developers and development times in literature supplied with each package of film. *Eastman Kodak Company*

tion, Figure 28-11. Then add the distilled water. Be certain to not overfill. Too much water will dilute the developer, causing it to be weak. The developing time is based upon a standard developer solution strength. Stir the developer thoroughly with a mixing rod.

STOP BATH. The stop bath working solution is used directly from the supply container. Pour the correct amount of solution into a graduate. Place it to the right side of the developer graduate.

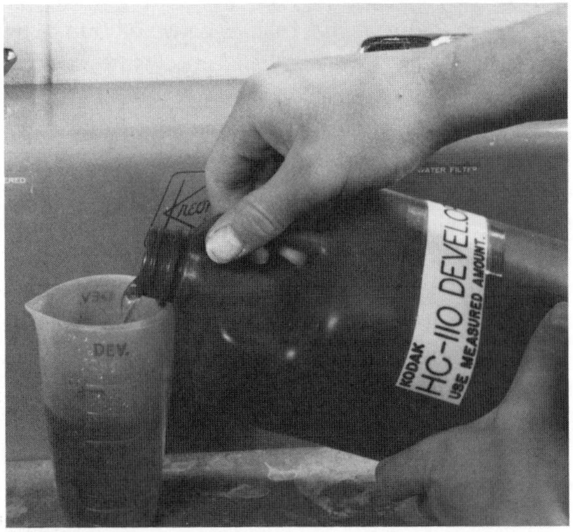

FIXER. Fixer also is used directly from the supply container. Because fixer can be reused several times, a used fixer container should be prepared. Fixer can be reused until it is exhausted. The dissolved silver that accumulates finally makes it ineffective. Exhausted fixer has a "muddy" appearance and loses its "grippy" feeling (feels slimy). Sometimes it even has residue on top. Position the fixer graduate filled with the correct amount to the right of the stop bath graduate.

The three chemical solutions that must be used in rapid order are now ready for use, Figure 28-12. It is a good habit to always work from left-

FIGURE 28-11. Pouring the correct amount of developer stock solution into a graduate.

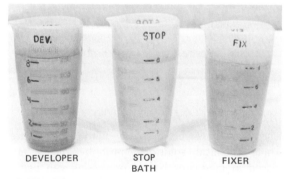

FIGURE 28-12. The three primary chemical solutions ready for use in processing film.

to-right with all photographic chemistry. Another good practice is to label the graduates with permanent markings. In this way, the graduates can always be used for the same chemical solution. This reduces the possibility of chemical contamination.

The remaining two chemical solutions, clearing agent and wetting agent, do not need pre-preparation. These two solutions can be used directly from the working solution containers. Further, there is no specific need to hurry when these processing steps are reached.

SUMMARY

Preparing the darkroom, loading the film into a tank, and preparing the chemistry are important steps. Quality film processing is impossible unless specific attention is given to preparation. If done poorly, any one of these steps could be the cause of substandard film processing.

REVIEW QUESTIONS

Answer these questions to test your knowledge of the unit content.

1. T/F Careful preparation for black-and-white film processing is of questionable value.
2. Why isn't it good practice to use an oil-based cleaner solution when cleaning a darkroom?
3. Name the four items that are needed when loading 35-mm film into a processing tank.
4. T/F It is wise to practice loading film into a processing tank with the light on.
5. When is the leader of an exposed roll of 35-mm film removed for processing?

   A. After exposure and while in the magazine
   B. Before removing film from the camera
   C. After removal from the magazine
   D. While in the processing tank

6. What condition should the processing tank reel be in before loading film into it?
7. T/F Tape used to hold film on a spindle should be completely removed before processing.
8. T/F Care must be exercised because roll film processing tanks may allow light to leak in even with the lid securely in place.
9. The optimum temperature of film processing chemistry is _____ °C (_____ °F).
10. How much chemistry is needed of each solution to process a single roll of film in a standard tank?

    A. 5 oz   C. 500 ml
    B. 250 ml   D. 12 oz

PROCESSING BLACK-AND-WHITE FILM

OBJECTIVES   *Upon completion of this unit, you will be able to:*
- *Develop a black-and-white roll of film.*
- *Stop and fix the film after it is fully developed.*
- *Wash and dry the film according to standard processing procedures.*
- *Finish the film for later use and storage.*

KEY TERMS   *The following new terms will be defined in this unit:*

Agitate   *Negative Sleeve*

Airbells   *Reticulation*

INTRODUCTION

Each and every step involved in processing black-and-white film must be done with care. Time, temperature, and agitation are ingredients to achieving quality. The importance of handling and storing film negatives is critical to producing photographic prints.

It is very important to use chemistry and water with a temperature at or very close to 68 °F (20 °C) throughout film processing. A sudden increase or decrease in temperature can cause *reticulation*. This condition can be recognized when film emulsion separates from the film base. The emulsion then cracks and peels away, leaving the area transparent.

FILM PROCESSING: DEVELOPING, STOPPING, FIXING, WASHING, AND DRYING

Complete these steps to process a 20- or 36-exposure, 35-mm roll of black-and-white film.

DEVELOPING

1. Determine the temperature of the developer solution. Use an accurate thermometer.

2. Using the temperature as a guide, look at a developing chart that has been prepared for the film being processed, Figure 29-1. As stated in unit 28, the optimum photographic chemistry temperature is 68 °F (20 °C). Other temperatures, plus or minus a few degrees, can be used successfully. Based upon the temperature and the kind/type of developer being used, determine the length of developing time. Different emulsion densities and contrasts can be obtained with specific developing times. Calculator tables and dials can be used to arrive at the correct development information. These are available directly from the film manufacturers.

3. Make certain that the loaded processing tank is very close to 68 °F (20 °C). It may be necessary to place the tank in a tray of 68 °F (20 °C) water to bring it up or down in temperature.

4. Set the timer according to the developing table in Figure 29-1.

5. Hold the tank in one hand, the graduate in the other, and pour the developer solution in as rapidly as possible, Figure 29-2.

6. Start the timer immediately after all developer is poured into the tank.

7. Tap the bottom of the tank on the sink or counter. This helps to remove *airbells* that may have formed when the developer was being poured in. Airbells are bubbles of air that form

KODAK Packaged Developers	DEVELOPING TIMES IN MINUTES*									
	SMALL TANK†—Agitation at 30-Second Intervals					LARGE TANK—Agitation at 1-Minute Intervals				
	65°F 18°C	**68°F 20°C**	70°F 21°C	72°F 22°C	75°F 24°C	65°F 18°C	**68°F 20°C**	70°F 21°C	72°F 22°C	75°F 24°C
HC-110 (Dilution B)	6	**5**	$4\frac{1}{2}$	4	$3\frac{1}{2}$	$6\frac{1}{2}$	**$5\frac{1}{2}$**	5	$4\frac{3}{4}$	4
POLYDOL	$6\frac{1}{2}$	**$5\frac{1}{2}$**	$4\frac{3}{4}$	$4\frac{1}{4}$	$3\frac{1}{4}$	$7\frac{1}{2}$	**6**	$5\frac{1}{2}$	$4\frac{3}{4}$	$3\frac{3}{4}$
D-76	$6\frac{1}{2}$	**$5\frac{1}{2}$**	5	$4\frac{1}{2}$	$3\frac{3}{4}$	$7\frac{1}{2}$	**$6\frac{1}{2}$**	6	$5\frac{1}{2}$	$4\frac{1}{2}$
D-76 (1:1)	8	**7**	$6\frac{1}{2}$	6	5	10	**9**	8	$7\frac{1}{2}$	7
MICRODOL-X	8	**7**	$6\frac{1}{2}$	6	$5\frac{1}{2}$	10	**9**	8	$7\frac{1}{2}$	7
MICRODOL-X (1:3)	—	**—**	11	10	$9\frac{1}{2}$	—	**—**	14	13	11

* Unsatisfactory uniformity may result with development times shorter than 5 minutes.

† Usually one-quart size or smaller.

Note: Do not use developers containing silver halide solvents.

FIGURE 29-1. A table that gives useful developing information. *Eastman Kodak Company*

FIGURE 29-2. Pour the developer solution into the processing tank as rapidly as possible.

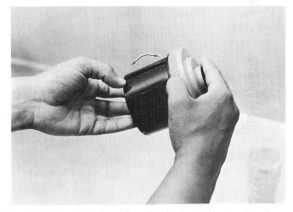

FIGURE 29-3. Agitating the developing solution in the processing tank.

on the surface of the film and cause uneven development.

8. Turn the tank upside-down two or three times and continue agitating for the first 30 seconds, Figure 29-3. *Agitate*, meaning move, the tank with a slow and uniform back-and-forth and side-to-side motion. Some processing tanks can be turned over without discharging the chemistry. This makes it possible to use an over-and-back agitating

motion. Proper agitation evenly distributes chemistry over the film. In this way, all areas of the film receive the same chemical treatment.

9. Set the tank down for 25 seconds and then agitate for 5 seconds.

10. Repeat step 9 until the last 30 seconds of development.

11. Begin pouring the developer out when there is 10 seconds of developing time left.

"One-time" developers should be properly discarded. Replenisher type developers should be poured into a properly labeled container.

STOPPING

12. Pour the prepared stop bath solution into the tank in the same rapid manner as was done with the developer.

13. Agitate the stop bath solution for the entire 30-second time period. It is not necessary to invert the tank as may have been done during development. A side-to-side agitation pattern is sufficient.

14. Discard the stop bath solution at the end of the 30-second time period.

FIXING

15. Set the timer for the specified amount of time. Regular strength fixer solution takes about 8 minutes to clear and harden the film. A stronger rapid fix solution needs only 3 to 4 minutes of working time.

16. Pour fixer solution into the processing tank. There is no special need to hurry at this point as the developing action was interrupted by the stop bath.

17. Agitate (side-to-side motion) the fixer for the first 30 seconds. Complete the fixing time period with the 5- and 25-second agitation-and-rest time cycle.

18. Pour the fixer solution in the "used-reuse" fixer container. Be careful not to dilute the fixer with water or contaminate it with other chemistry.

WASHING

19. Remove the lid from the processing tank. The film is no longer sensitive to light.

20. Give the film a 30-second water rinse. Fill the tank with tap water, and agitate continuously for the entire time period. Pour out the water.

21. Fill tank with a premixed hypo clearing agent. Provide moderate but continuous agitation for 2 minutes.

22. Pour the clearing agent back into its regular storage container. It can be used many times.

23. Wash film for 15 minutes with running water. Place a hose in the ratchet core and run water so it moderately overflows the tank, Figure 29-4.

24. Pour out some of the water to permit the addition of the wetting agent. Only two capfuls or approximately 1 oz (31 ml) need to be added. Agitate continuously for 30 seconds.

DRYING

25. Lift the ratchet and film out of the tank. Carefully remove the film, now negatives, from the ratchet.

WATER HOSE

FIGURE 29-4. Washing film to remove the processing chemistry.

26. Attach a film clip to each end of the film. Be certain to hold the film negatives over the sink so that the water drips into the sink instead of onto the floor, Figure 29-5.

27. Remove the excess water from the film. It is best to use a soft blade squeegee, sponge, or chamois that has been dipped into the wetting agent. Wipe over the film with care so that the soft emulsion is not scratched.

28. Hang the film negatives in a dust-free area to dry. Use a drying cabinet if available. Drying time can vary from 30 minutes to 2 hours depending on the conditions.

Clean up the darkroom. Place all chemistry containers in their proper locations. Wash and dry the entire processing tank, graduates, thermometer, mixing rods, and other tools used to process the film. Clean the sink and work counter. Leave the darkroom in the best condition possible.

FINISHING

Cut the film in specific exposure lengths. A negative strip should contain five or six exposure frames. This provides a strip that is long enough to handle conveniently in an enlarger. Cut very carefully between two exposure frames. Use a sharp scissors and handle the negatives by the edges, Figure 29-6.

Slide the negative strips into protective sleeves of plastic or translucent paper. Sleeves are available that hold single or multiple negative film strips. The type used is not important. Using *negative sleeves* to protect film strips is important.

FIGURE 29-5. Film clip attached to the processed film negatives.

FILM CLIP

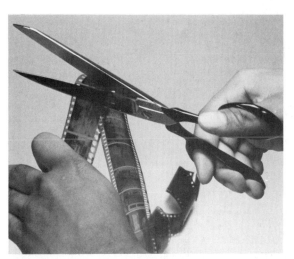

FIGURE 29-6. Cutting a roll of processed film into negative strips that contain 5 or 6 exposures.

Filing individual, large-size negatives and smaller size negative strips has distinct value. Many commercial systems are available, such as the one shown in Figure 29-7. This system uses negative sleeves, envelopes for the sleeves, and filing boxes. Room is provided on the envelopes to identify the negatives. Also, a card with several lines is located in the box lid where information can be easily recorded.

Store film negatives in a cool and dry area. Heat and moisture are enemies of photographic film both before and after processing. Also, make certain that the storage area is free of dust and dirt. Keep lids on file boxes and seal them with tape during long storage periods.

ANALYZING BLACK-AND-WHITE NEGATIVES

Inspect the negative film strip after it is completely dry. Lay the film strip, emulsion side down, on a light table, Figure 29-8. First, inspect the entire negative film strip for flaws or problems. Next, use the magnifying glass and closely inspect the obvious flaws. Make every attempt to determine the problem, its possible cause, and how to correct it.

This information will be valuable for future film processing.

Review the entire 20 or 36 negatives with the magnifying glass. Problems may be present that go undetected until enlargements from the selected negatives are made. After a problem has been identified, the cause must be determined. A solution is then possible, Figure 29-9. Remember—"Problems never fix themselves."

PUSH PROCESSING

Film that has been exposed at twice, four, or more times its rated speed must receive special developing attention. Generally, 50% more developing time is required each time the speed of the film is doubled. For example, ISO 400 film is exposed as if it had a rating of ISO 800. If the developing time for the ISO 400 rating was 5 minutes, then the increased developing would require an extra $2\frac{1}{2}$ minutes or a total of $7\frac{1}{2}$ minutes.

Pushing film to speeds higher than the manufacturer's rating has some disadvantages. Shadow detail is often reduced or lost entirely. Also, the negatives are grainier and less sharp. The major advantage of "pushing" film is to permit taking pictures in low light without using a flash. The photographer must decide if the reduced quality is worth the need to take pictures in less than fully lighted areas.

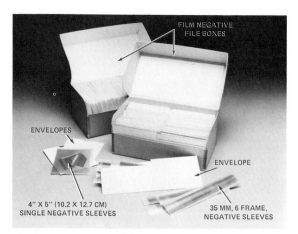

FIGURE 29-7. A negative storage system that includes sleeves, envelopes, and file boxes. *Light Impressions Corporation*

FIGURE 29-8. A light table is useful when inspecting processed negatives.

ANALYZING BLACK-AND-WHITE NEGATIVES		
Problem	**Cause**	**Solution**
Too dark or totally black	Film overexposed or developed too long. Stop bath or fixer exhausted	Check camera TTL light meter and shutter operation. OR Check processing instructions and prepare new chemistry.
Too light or totally clear	Film underexposed, developer solution weak, or development length too short	Check camera TTL light meter and shutter operation. OR Use new developer solution and confirm development length.
Scratches in the film emulsion	Dirt and other foreign particles on the squeegee blades, sponge, or chamois when removing water from the wet film, or striking the wet (soft emulsion) film against the sink, cabinet, or film dryer	Keep the squeegee, sponge, or chamois clean. Wipe over the soft-emulsion with utmost care. Handle the film carefully at all times.
Clear or near-clear spots on the negatives	Bubbles of air, called "Airbells," that adhered to the film. The bubbles did not permit developer solution to reach the film	Tap the processing tank sharply on a hard surface at the beginning and even during development.
Film emulsion has a crinkled surface similar to the surface of an orange	This is called reticulation. Sudden temperature change in the processing chemistry. Also, major differences in the acidity or alkalinity of the processing solutions	Maintain a constant temperature for all of the chemical solutions. Mix all chemical solutions carefully to maintain a balance in the acidity and alkalinity.
Uneven development of part or all of the roll of film	Not enough chemistry to cover the film during development, or film touching causing noncirculation of the chemistry	Use enough chemistry to completely cover the film, and load the film properly in the processing tank so the chemistry can circulate freely on all portions of the film.
"Teardrop" shaped marks on the surface of the processed film	Accumulation of water after the film has been squeegeed that leaves deposits of material while the film dries.	Squeegee the film carefully and completely remove all wash water before drying. If teardrop marks appear, carefully remove them by wiping the film with a lint-free damp cloth. Sometimes it is necessary to completely rewash and dry the film.

FIGURE 29-9. A problem-cause-solution chart is often useful in analyzing processed black-and-white negatives.

SUMMARY

Film processing includes developing, stopping, fixing, washing, and finishing. Each of these stages is equally important to the finished product. Errors in any one of the total series of steps can make the film useless. Set up a standard operations procedure (SOP) and always do the processing exactly the same way each time. Exercise care with even the smallest of details to achieve quality black-and-white film negatives. After processing, black-and-white negatives should be analyzed thoroughly.

REVIEW QUESTIONS

Answer these questions to test your knowledge of the unit content.

1. What is the most frequent cause of reticulation?
2. T/F More developing time is needed when the developer solution is colder than the optimum processing temperature.
3. Developer solution should be poured— slowly or rapidly—into the tank.
4. Following the initial 30 seconds, agitation for both the developer and fixer should be done for 5 seconds every:

   A. 25 seconds   C. 45 seconds
   B. 15 seconds   D. 35 seconds

5. T/F Stop bath solution should receive continued agitation during its contact with the film.
6. Of the five standard film processing solutions, name the two that are saved and reused.
7. When can the lid of the processing tank first be removed during film processing?
8. For drying purposes, a 35-mm roll of 20- or 36-exposure film should have _____ film clips attached to it.
9. A 35-mm roll of processed film should be cut into strips of how many frames each?

   A. 2 to 3   C. 4 to 5
   B. 3 to 4   D. 5 to 6

10. When pushing film, how much more developing time is generally recommended for each doubling of the film speed?

    A. 10%   C. 50%
    B. 25%   D. 75%

CHAPTER

SIX

PRINT PROCESSING—BLACK-AND-WHITE

PHOTOGRAPHIC PAPER AND IMAGING

OBJECTIVES   *Upon completion of this unit, you will be able to:*
- *Sketch the layer structure of both fiber-based and resin-coated photographic paper.*
- *Explain some critical aspects of forming the image in photographic paper.*
- *List six important characteristics of photographic paper.*
- *Describe how photographic paper prints were first made on a mass production basis.*

KEY TERMS   *The following new terms will be defined in this unit:*

Baryta Coating	Silver Bromide	Tone
Fiber-based	Silver Chloride	Variable-contrast
Resin-coated	Silver Chlorobromide	Weight

INTRODUCTION

Photographic paper is a product that faithfully reverses negative film images. It is so sensitive that it is possible to record even the smallest image areas. A pin or needle point is not too small. Technically, photographic paper should be called "continuous-tone" paper. Its emulsion, just as with continuous-tone film, is capable of recording the complete gray scale from the lightest to the darkest images.

STRUCTURE OF PHOTOGRAPHIC PAPER

Photographic paper is made of layers similar to photographic film. The number of layers and contents of each depend upon the kind and type of paper. Basically, there are two types of papers used to make photographic prints: fiber-based paper and resin-coated paper.

FIBER-BASED PAPER.   Of the two types, this is by far the oldest type of paper. It has been in existence since photography was available to the general public in the late 1800s. Fiber-based paper contains four basic layers, Figure 30-1. The *protective coating* is made of gelatin and protects the emulsion from abrasive scratches. Without this coating, the

emulsion could easily become scratched during packaging at the factory, while in the package, or during use in the darkroom.

The *emulsion layer* consists of minute particles of silver halide crystals suspended in gelatin. The light-sensitive substance used in the emulsion depends upon the end use of the paper. The mixture is much like that used in continuous-tone film except that it is less light-sensitive. *Brightness coatings* are most often made from barium sulfate; thus, the term *Baryta coating* is frequently used. The photographic print reflects more light with such a coating

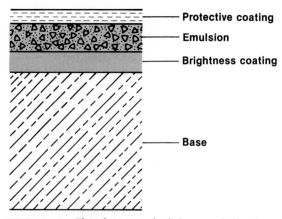

FIGURE 30-1. The four standard layers of fiber-based photographic paper.

under the emulsion. This gives the finished print a sharp and crisp look.

The *base* of photographic paper is made from wood pulp, just as with writing and regular printing papers. Paper for the base must be consistent in thickness and in strength. Dimensional stability contributes significantly to the quality of the finished photographic print.

RESIN-COATED PAPER. The major difference of resin-coated (RC) paper from fiber-based paper is the application of two layers of plastic resin, Figure 30-2. Coating both sides of the paper base makes the photographic paper water-resistant. Because of this feature, processing time is reduced considerably.

FORMING THE IMAGE

Most black-and-white photographic papers are orthochromatic, making them sensitive only to blue and weak-side green wavelengths of visible light, Figure 30-3. This makes it possible to use darkroom safelights that fall between green and red (yellow)

and completely red. It is necessary to use panchromatic paper when black-and-white prints are made from color negatives. This permits all three primary wavelengths of light to work together in forming an image in the silver halides of the paper emulsion.

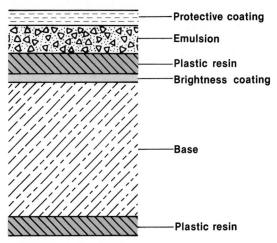

FIGURE 30-2. The six standard layers of resin-coated photographic paper.

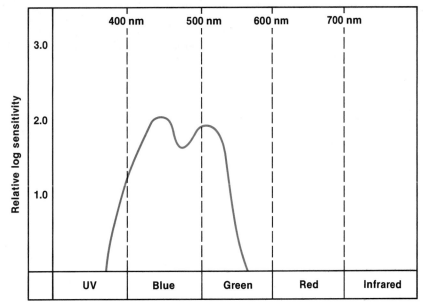

FIGURE 30-3. A wedge spectrogram illustrates the light sensitivity of orthochromatic printing paper.

When using panchromatic paper, it is essential to work in total darkness.

Emulsions vary depending on the imaging needs. *Silver chloride* emulsions are very slow and are used in papers designed for contact printing. This is when the negative and paper are in direct contact and a bright light source will be used to expose the image. *Silver bromide* emulsions are much faster and are used for projection papers. These papers are most often exposed with enlargers, which are not capable of producing enough light to expose slow emulsions.

Most photographic papers contain a combination of silver chloride and silver bromide crystals. This combined mixture is called *silver chlorobromide*. The combination of the crystals provides for a greater control of the emulsion sensitivity to light. Projection papers often are used for contact printing with an enlarger as a light source. Direct light from contact printing units is too close and intense for projection papers.

Latent images are formed in paper emulsions in the same manner as with film. They only become visible when contact is made with the liquid developing solution. As with film, paper latent images are entirely unseen, but once they are part of the emulsion, they will remain for long periods of time. The type of emulsion, length of exposure, and storage conditions are variables affecting the length of time an exposed sheet of paper can be saved before processing. As with most photographic materials, it is best to process immediately following exposure.

CHARACTERISTICS OF PHOTOGRAPHIC PAPER

There are a number of characteristics that can be used to identify photographic papers. Among them are image surface, contrast, tone, speed, weight, and size.

IMAGE SURFACE. The most popular surface used is that which produces a glossy finish. Other surface textures include high luster, luster, and matte. The difference in the surface when compared to glossy is that each is less shiny. Special surface textures such as silk, tweed, and canvas are useful for finished prints such as portraits.

CONTRAST. Two categories of papers relate to contrast—graded and variable-contrast. This paper characteristic permits the photographer to control the contrast between the highlights and shadows. *Graded* papers are those designed to produce a specific range of tonal values between the lightest and darkest areas of the image, Figure 30-4. The grades range from 1 through 5, or are identified as L, M, and H. Number 1 and L graded papers are of low contrast and should be used with high contrast negatives. Number 3 and M papers are of medium contrast. Number 5 and H papers are of high contrast and are used with low contrast negatives.

Variable-contrast papers make it convenient to control the range of tonal values in a negative image. These papers contain a mixture of two color-sensitive emulsions. One emulsion is sensitive to green wavelengths, while the other emulsion is sensitive to the blue-violet wavelengths of light. The green-sensitive emulsion produces low-contrast images with many shades of gray. The blue-violet emulsion produces high-contrast images with intense blacks.

Special variable-contrast filters are used in the enlarger when these papers are exposed. The color of the exposing light is changed by selecting different filters. Selecting a filter that transmits a majority of green light affects the low-contrast emulsion. High-contrast images are obtained by projecting blue–violet light, thereby affecting the emulsion that produces the intense blacks. Use of these filters is described in unit 35.

TONE. Emulsions are formulated to produce tonal values that vary from brownish-black to bluish-black. Those that tend to appear on the brown side are considered "warm" papers. Blue-black appearing images are identified as "cold"

LOW CONTRAST PRINT	MEDIUM CONTRAST PRINT	HIGH CONTRAST PRINT
PAPER GRADE L	PAPER GRADE M	PAPER GRADE H

ALL THREE PRINTS MADE FROM THE SAME NEGATIVE.

FIGURE 30-4. Graded papers are useful in changing the range of tonal values in the finished print. *Eastman Kodak Company*

papers. Paper developers are available that increase the warm or cold image appearances.

SPEED. Photographic paper light sensitivity is measured in general terms. Each type of paper is simply listed as being medium, medium-fast, or fast speed. A numerical rating system for photographic papers does not exist as there does with film.

WEIGHT. Weight refers to the thickness of the paper base upon which the several coatings

and layers are placed. Photographic papers are available in light weight (LW), single weight (SW), medium weight (MW), and double weight (DW), Figure 30-5. Medium weight papers are used most frequently for general purpose photography.

SIZE. Photographic papers are available in standard size rolls and sheets. Precut sheets are packaged most often in 25, 100, and 250 sheet amounts. Sheet sizes range from $2\frac{1}{2}'' \times 3\frac{1}{2}''$ (6.4 to 8.9 cm) to $20'' \times 24''$ (50.8 × 61 cm). Special roll

Paper	Approximate Thickness
Light-weight	4.50 mm
Single-weight	6.75 mm
Medium-weight	9.00 mm
Double-weight	14.00 mm

FIGURE 30-5. The common weight designations of photographic paper and their associated millimeter thicknesses.

and sheet sizes are available directly from the manufacturer. For this service, large quantities must be purchased at one time.

HISTORICAL HIGHLIGHTS

The first photographs were created on metal and glass surfaces. This was due to the nature of creating images. When flexible film became available during the 1880s, lightweight photographic papers then became a reality. In 1883, The Eastman Dry Plate and Film Company (now Eastman Kodak Company) moved into its new building, Figure 30-6. The area that looks like a greenhouse on top of the building served as the printing room. It was here that negatives were placed in contact with the printing paper, Figure 30-7. Sunlight was used for exposure. After exposure, the paper was transferred to a fixing solution to be made permanent.

SUMMARY

There are many brands and types of photographic paper. With these papers come a variety of characteristics, speeds, and weights. The choices seem almost endless; thus, photographers and consumers should be able to find the right paper for any photographic need.

REVIEW QUESTIONS

Answer these questions to test your knowledge of the unit content.

1. Because of its capability, photographic paper could actually be called _____ paper.

FIGURE 30-6. The 1883 factory of the Eastman Dry Plate and Film Company (now called Eastman Kodak Company). *Eastman Kodak Company*

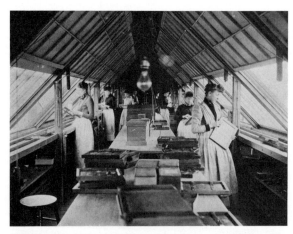

FIGURE 30-7. Photographic prints were exposed by sunlight in the 1894 photofinishing operation of the Eastman Company. *Eastman Kodak Company*

2. Why is a coating of transparent gelatin placed on top of the emulsion of photographic paper?

3. T/F The brightness coating is placed directly under the emulsion in both fiber-based and resin-coated photographic papers.

4. T/F Orthochromatic photographic papers are safe from being exposed by an approved red safelight.

5. Which compound used in making emulsions for photographic paper has the greatest sensitivity to light?

A. Silver bromide
B. Silver halide
C. Silver chloride
D. Silver chlorobromide

6. T/F Photographic papers are capable of retaining latent images for only a few minutes.

7. The most popular image surface for photographic papers is:

A. High luster   C. Luster
B. Matte   D. Glossy

8. What colors of light-sensitive emulsions are used to make variable-contrast photographic papers?

9. What does the term "weight" mean with reference to photographic paper?

10. Name the light source used to expose photographic prints in the Eastman Kodak factory prior to the year 1900.

UNIT 31

FACILITIES FOR PRINT PROCESSING

OBJECTIVES   *Upon completion of this unit, you will be able to:*
- *Specify equipment needed for print processing.*
- *Assemble the various tools useful in exposing and processing photographic prints.*
- *Prepare a floor plan of a basic print processing darkroom.*

KEY TERMS   *The following new terms will be defined in this unit:*

Easel   Focusing Aid
Enlarger   Print Tongs
Ferrotype

INTRODUCTION

The best type of print processing facilities are those separated from film processing facilities. Some aspects of film and paper processing are different; thus, it could be confusing if both are done in the same processing sink. This is especially true of the chemistry. If both film and paper must be processed in the same darkroom, make sure chemistry containers are clearly identified.

EQUIPMENT

A minimum of 12 pieces of equipment are needed in a print processing darkroom. Each of these is described in the following paragraphs and figures. In addition, an electronic air cleaner helps to ensure a dust-free environment.

    ENLARGER. An *enlarger* is basically a simple device designed to shine light through a film negative and expose photographic paper, Figure 31-1.

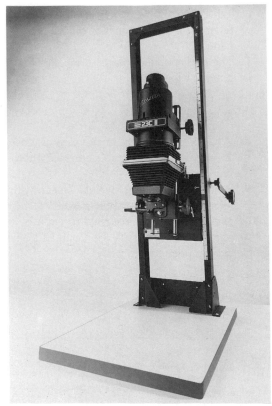

FIGURE 31-1. A solidly built enlarger is the heart of a print processing darkroom. *Charles Beseler Company*

FIGURE 31-2. Easels hold photographic paper secure during the exposure. *Doran Enterprises, Inc.*

It must be capable of evenly illuminating the film so that the image is faithfully projected onto the paper. Further, the entire unit must be solidly built. There must be no vibration of the exposing head during an exposure.

An enlarger must be equipped with a quality lens if sharp images are expected. A 50-mm lens is standard equipment on an enlarger that will be used primarily for 35-mm film. Lenses with shorter focal lengths give greater magnifications, making them good for film that is smaller than 35 mm. Lenses with longer focal lengths decrease the magnification factor; thus, they should be used for larger-sized negatives. Image magnification varies from 8 times (8 ×) to 33 times (33 ×) with different brands and types of enlargers equipped with a 50-mm lens.

EASEL. An *easel* holds photographic paper in position while paper is exposed, Figure 31-2. Easels are available in a wide variety of sizes and styles. Most are designed to create borders on the photographs, but several permit "borderless" prints. Some easels accept one or more standard sizes of cut paper. Others are adjustable, making them capable of accepting nearly any paper size within their design limitations.

TIMER. It is essential to have a timer hooked into the electrical system of the enlarger, Figure 31-3. Timers must permit accurate exposure control on a repetitive basis to be most effective. All controls and time settings should be easily visible in darkroom lighting.

FOCUSING AID. This is a useful device that magnifies the image being projected by an enlarger, Figure 31-4. It is placed on the easel directly below the lens. The enlarger can be quickly and accurately focused by viewing the magnified projected image.

TRAYS. High-impact resistant plastic or stainless steel trays are essential for processing photographic prints. Standard sizes range from 5″ × 7″ (12.7 × 17.8 cm) to 20″ × 24″ (50.8 × 61 cm). Their depths vary depending on the outside dimensions. Smaller trays are about 2″ (5.1 cm) deep, while the larger trays are up to 6″ (15.2 cm) deep. A minimum of three trays will always be needed.

SINK. A sink should be large enough to conveniently hold the three processing trays plus a washing tray or system, Figure 31-5. High-impact plastic or stainless steel sinks provide the best service. A good sink design includes shelving for chemical storage under the sink area.

PRINT WASHER. Many types of print washers are available, Figure 31-6. The kind or type used is not important. What is important is that the prints are thoroughly washed with circulating water in the shortest period of time possible.

PRINT DRYER. These units can vary from a roll of blotter paper to a high-speed variable heat power unit, Figure 31-7. Resin-coated papers cannot withstand high heat, thus care must be taken when drying these papers. It is best to "air dry" resin-coated papers whenever possible. Fiber-based papers need considerable heat for drying, and a unit containing one or more ferrotype plates is useful, Figure 31-8. A *ferrotype* plate is a highly polished

FIGURE 31-3. A darkroom timer designed to control exposures accurately on an enlarger. *Dimco-Gray Company*

FIGURE 31-4. A focusing aid helps to focus an enlarger quickly and accurately. *Bestwell Optical Instrument Corp.*

FIGURE 31-5. A high-impact fiberglass sink on a solid frame works very well for photographic print processing. *Omega Arkay*

FIGURE 31-6. A print washer must circulate water over photographic prints in a complete manner. *Doran Enterprises, Inc.*

FIGURE 31-7. A high-production heated dryer for resin-coated (RC) papers. *Omega Arkay*

FIGURE 31-8. A high-heat ferrotype dryer for fiber-based papers. *Doran Enterprises, Inc.*

FIGURE 31-9. A photographic processing chemistry bulk container. *Omega Arkay*

sheet of chrome-plated steel that helps add a glossy surface to fiber-based papers. A ferrotype dryer may or may not be heated.

TRIMMER. A bar-type or wheel-type paper trimmer is very useful. There is need on occasion to cut photographic paper for specific purposes.

CHEMISTRY CONTAINERS. Airtight, plastic containers for mixed chemistry are essential, Figure 31-9. A good type is one with a valve that permits easy drawing of chemistry directly into a tray.

FIGURE 31-10. One or more safelights are essential in a photographic darkroom. *Doran Enterprises, Inc.*

FIGURE 31-11. A light safe keeps photographic paper from being fogged accidently. *Doran Enterprises, Inc.*

FIGURE 31-12. An electronic air cleaner is helpful in keeping a darkroom free of airborne dust particles.

SAFELIGHTS. One or more safelights must be located within a darkroom, Figure 31-10. For orthochromatic photographic paper, red or yellow safelights can be used. It is important to use safelights that are commercially approved and tested. See the manufacturer's recommendation on the package of paper. A standard red or yellow light bulb purchased at a local retail store may not be "safe," meaning it may fog (slightly expose) the photographic paper.

PAPER SAFE. A light-tight cabinet is needed to protect the supply of photographic paper from being exposed, Figure 31-11. A cabinet should be easy to open and close, but it must be 100% light proof. Those with shelves permit several sizes of paper to be available at any one time.

AIR CLEANER. Dust and dirt must not be permitted in a print processing darkroom! An electronic air cleaner helps keep darkroom air clean of odors and small particles of dust, Figure 31-12. The financial investment for an air cleaner device or system will be paid back many times over with dust-free printing problems.

TOOLS

Some special tools are essential, while others are nice to have in a darkroom. The following tools are those which can be considered essential. As experience is gained, more tools will be identified and acquired.

PRINT TONGS. These are "finger-like" devices that help in handling wet prints. Fingers should not be used to transfer wet prints from one chemical tray to another. Plastic or metal print tongs can be used to do the job much better.

SQUEEGEE. This blade-like device is useful in removing water from the face of a print, Figure 31-13. The rubber or plastic blade must not scratch the soft emulsion. A *chamois* works very well to remove surface water from a resin-coated print. Sometimes this is better than a regular squeegee.

ROLLER. A single or double roller is needed to press fiber-based prints onto ferrotype plates, Figure 31-14.

FIGURE 31-13. A squeegee is used to remove water from a washed print. *Jobo Fototechnic, Inc.*

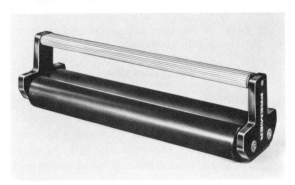

FIGURE 31-14. A double roller used to press wet prints on ferrotype plates. *Doran Enterprises, Inc.*

NEGATIVE BRUSH OR FILM CLEANING MACHINE

A soft brush charged with low radioactive particles helps remove dust and dirt from negatives, camera lenses, and enlarger lenses. The brush works by ionizing the air in the immediate vicinity of where it is being used. As do brushes, electronic film cleaners work quietly and efficiently through the ionization of the film surfaces, Figure 31-15. Pressure containers of air also work well to clean photographic products and equipment.

FIGURE 31-15. A film cleaning machine that ionizes the film surfaces as the film is drawn through the brushes. *Kinetronics Corporation*

DARKROOM

Print processing darkrooms should be designed for convenient access and traffic flow. This is true if the darkroom will be used by one or dozens of people. A "one enlarger" darkroom is illustrated in Figure 31-16. It provides an ideal working environment for one and possibly two people at a time.

A darkroom should have a convenient entrance and exit light-trap system. This permits people to enter and leave without affecting the darkroom lighting conditions. A good darkroom must have two specific areas. These are the dry area where the paper is exposed and the wet area where it is processed. The designation of these two areas is essential for quality work.

Storage areas, such as shelving and drawers, are useful in any darkroom. This is especially true

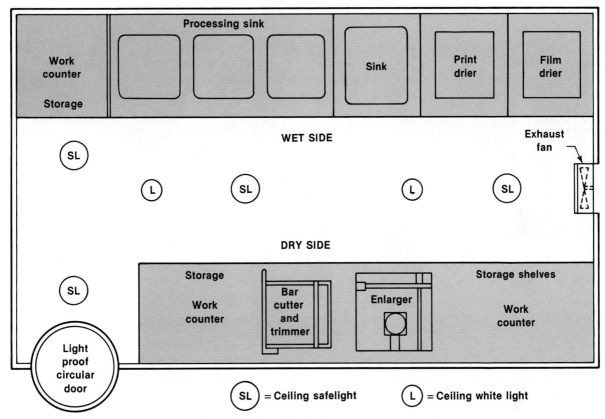

Work counter	Processing sink			Sink	Print drier	Film drier

Storage

WET SIDE

Exhaust fan →

SL

L   SL   L   SL

DRY SIDE

SL

Storage   Storage shelves

Work counter   Bar cutter and trimmer   Enlarger   Work counter

Light proof circular door

SL = Ceiling safelight   L = Ceiling white light

FIGURE 31-16. Floor plan of an ideal one to two person darkroom.

in darkrooms designed for use by many people, Figure 31-17. Chemical and paper storage areas should be accessible to both short and tall people. Storage drawers for work-in-progress can save considerable time.

Darkrooms need not have solid walls made of wood or concrete products. They simply need to be made free of white light and have necessary work flow features. Walls should be painted a light color so that the safelight will reflect from them. This makes the darkroom lighter but still safe for the photographic paper. Dark walls absorb too much light, making it more difficult to see and work. Portable darkrooms are available, making it possible to establish a "working darkroom" almost any-

FIGURE 31-17. A print processing darkroom designed and equipped for use by many people at one time. *West Virginia University, School of Journalism*

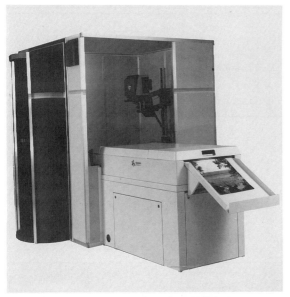

FIGURE 31-18. A compact, prefabricated darkroom containing an enlarger, a paper processor, a paper safe cabinet, and a revolving entry-exit door. *ESECO Speedmaster*

where, Figure 31-18. Manufacturing plants sometimes have need for facilities of this type.

SUMMARY

Print processing facilities must include certain equipment and tools. The number of each will be based upon the number of people who will use the darkroom at any one time. As might be assumed, several people can use the same equipment and tools on a rotation basis. It is critical that the same number of enlargers and associated equipment be available as there will be persons working in a print processing darkroom at one time. A major consideration in any darkroom is **cleanliness**, Figure 31-19. Everyone who makes use of a darkroom must be responsible for helping to **keep it clean**.

FIGURE 31-19. Wiping dust and dirt from an enlarger. Everyone must work to keep a darkroom clean. *MWB Industries, Inc.*

REVIEW QUESTIONS

Answer these questions to test your knowledge of the unit content.

1. T/F The ideal print processing darkroom is separate from the film processing darkroom.
2. The most appropriate lens to use in an enlarger that will be used primarily for 35-mm film is:

   A. 28 mm   C. 50 mm
   B. 35 mm   D. 75 mm

3. Name the device used to hold photographic paper in place while it is being exposed on an enlarger.
4. Why should a timer be hooked into the electrical system of an enlarger?
5. What are the two types of photographic print dryers?
6. T/F A red light bulb purchased from the local hardware store can safely be used as a safelight in a print processing darkroom.
7. What value is there in using an electronic air cleaner in a darkroom?

8. A brush capable of ionizing the air around film negatives contains _____ particles.

9. A good print processing darkroom will have a designated _____ area and a _____ area.

10. A print processing darkroom designed for four people to work at one time should contain how many enlargers?

A. 1   C. 3
B. 2   D. 4

— UNIT 32 —

PROJECTION PRINTING FUNDAMENTALS

OBJECTIVES   *Upon completion of this unit, you will be able to:*
- *Name the major parts of an enlarger.*
- *Explain the major differences of the three types of enlargers.*
- *State the distinguishing characteristics of the four light sources used in enlargers.*
- *Operate an enlarger to make test prints and enlargements.*

KEY TERMS   *The following new terms will be defined in this unit:*

Condenser Enlarger	Enlarger Head	Negative Carrier
Diffusion Enlarger	Highlights	Tonal Values

INTRODUCTION

Projection printing makes it possible to create an enlarged image from a small-sized negative. In concept, the procedure is simple. In practice, the procedure must be carried out with precision. Also, using quality equipment and materials are strong contributors to achieving successful photographic enlargements.

ENLARGER PARTS

The parts and controls of enlargers are important to learn, Figure 32-1. Each brand and model is designed a little different. This should not create a problem in operating any enlarger if the concepts and fundamentals are known and understood.

    SUPPORT FRAME. A strong support frame is essential to securely hold the projection portion of the enlarger. The major support is often a single circular post. Two rectangular-shaped posts are used on many of the heavy-duty enlargers.

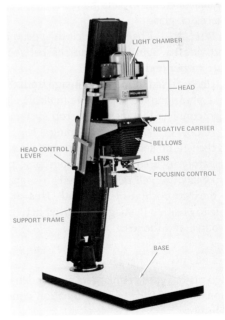

FIGURE 32-1. The basic parts of an enlarger. *Omega Arkay*

BASE. The platform on which the enlarger stands. It is most often made of wood. The base must be of sufficient size and thickness to give the enlarger a secure foundation.

HEAD. Includes the combined parts of the light chamber, negative carrier, bellows, and lens. It is the projection system of the enlarger.

LIGHT CHAMBER. Bright and consistent illumination is needed to project an image to the photographic paper. Different types of light and methods of directing it through the negative path are used on enlargers.

HEIGHT-ADJUSTING CONTROLS. The mechanism used to raise, lower, and lock the total enlarger head in position. The head must move up and down the support frame to change the projected image size. The head should move smoothly and be locked easily.

NEGATIVE CARRIER. The sandwich-type device that securely holds the negative in place while the exposure is made. Some negative carriers are designed to hold the film around the edges of a rectangular opening. Others hold the film between two pieces of glass.

BELLOWS. The flexible cloth-type material that provides a light-tight chamber between the negative carrier and the lens.

LENS. Used to focus the image from the film onto the photographic paper. The multiple element lens also contains the aperture/f-stop system.

FOCUSING CONTROL. Turning the knob moves the lens up and down. This permits the image to be sharply focused on the photographic paper.

HEAD CONTROL LEVER. Used to raise and lower the top portion of the head. This makes it possible to position the negative carrier between the light source and the lens.

ENLARGER TYPES

The two major types of enlargers are condenser and diffusion. A third type is a combination condenser and diffusion.

CONDENSER. The illumination system includes a light source and two or more bowl-shaped

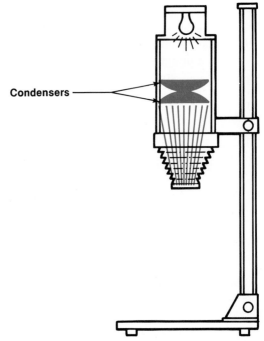

FIGURE 32-2. A condenser enlarger focuses light toward the negative.

glass lenses called condensers, Figure 32-2. Most enlargers used for black-and-white work are of the condenser type. The light rays are collected and focused through the negative to the lens and onto the print paper. The condensing of the light gives the resulting photographic prints a crisp and sharp appearance. The intensity of the light tends to emphasize the dust and dirt problems. This sometimes is a cause for concern.

DIFFUSION. The illumination system includes a light source, a mixing box, and a piece of frosted glass or plastic, Figure 32-3. This enlarger head design permits the light rays to blend together. The light strikes the negative from many directions. This condition gives a softer image on the photographic paper. Diffusion enlargers are most often used with larger sized negatives. Portrait prints are best made on a diffusion type enlarger as are other less-contrasty scenes.

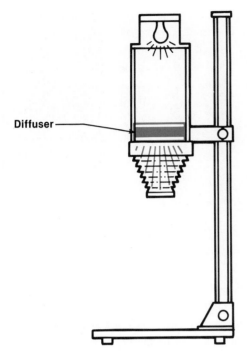

FIGURE 32-3. The diffusion enlarger distributes light evenly over the negative.

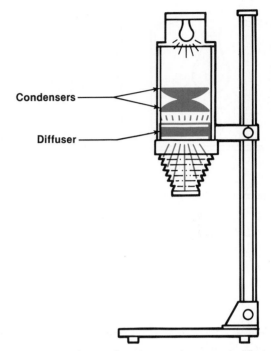

FIGURE 32-4. The combination condenser and diffusion enlarger provides a compromise.

COMBINATION. The condenser and diffusion combination type enlargers include both bowl-shaped condenser lenses and a piece of frosted glass, Figure 32-4. These enlargers provide a compromise of focused and diffused light passing through the negative.

ENLARGER LIGHT SOURCES

Artificial light is created from many different sources. Four categories of electric artificial light are used to transmit the image in the film to the photographic paper. These are incandescent, point source, cold-light, and high intensity lamps, Figure 32-5.

INCANDESCENT. These lamps contain a filament housed within a gas-filled or vacuum chamber. When charged with electrical current, the filament gives off a bright light. Incandescent lamps (bulbs) give off considerable heat. This sometimes causes the

negative to buckle. This condition is called "negative pop," and it shifts the negative out of focus. Small enlargements are not noticeably affected by negative pop, but larger prints of 8" × 10" (20.3 × 25.4 cm) and up are affected. The solutions to this problem are to work rather rapidly or use a negative carrier containing glass or use a larger lens opening.

POINT SOURCE. This special small bulb produces directional light. A transformer is used to control the intensity of the lamp while the aperture is left fully open. The bulb can be moved up and down in the lamp housing giving full directional control of the light rays.

COLD LIGHT. A fluorescent tube is located in the top portion of the lamp housing. The lamp creates very little heat and produces extremely soft and even light. This illumination source is most often used on diffusion type enlargers. It does require a warm-up before use.

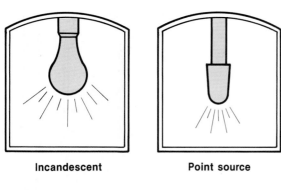

Incandescent   **Point source**

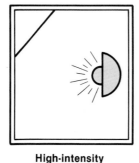

Cold-light   **High-intensity**

FIGURE 32-5. The four common light sources used in enlargers.

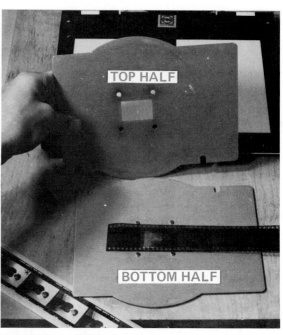

FIGURE 32-6. Placing the negative strip, emulsion down, in the negative carrier.

HIGH INTENSITY. Lamps of this type create extremely bright light. They also produce high levels of heat. A reflector is used to direct the light, thus making it possible to use all of the produced energy. High-intensity lighting systems are often used in enlargers specifically designed for color printing (see unit 39).

BASIC ENLARGING PROCEDURE

Making a projection print is an exciting venture. This is true whether it is the first enlargement ever made or whether a thousand prints have been produced over several years. The following steps will serve as a guide to make an enlarged photographic print.

1. Establish safelight conditions in the darkroom.

2. Obtain the negative strip containing the selected frame to be enlarged. Handle the negative strip by the edges.

3. Place the negative in the negative carrier with the emulsion down, Figure 32-6. Also, make certain the image on the negative is upside-down. In this position, the image will be right reading on the easel.

4. Hold the negative carrier so that light from the safelight permits exact positioning of the negative frame in the rectangular opening. Be careful to not scratch the negatives at this point.

5. Clean both sides of the negative with an anti-static brush. It is important to remove all dust and dirt from the surfaces of the selected negative frame.

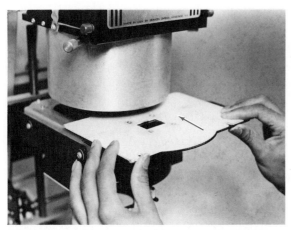

FIGURE 32-7. Positioning the negative carrier in the enlarger.

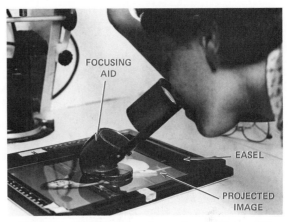

FIGURE 32-8. Fine focusing the projected image with a focusing aid.

6. Place the negative carrier in the enlarger, Figure 32-7. Make certain the carrier is positioned exactly against the guides. A mispositioned carrier may permit light to leak out and fog the photographic paper.

7. Turn on the enlarger light and open the lens to its maximum aperture. This makes it possible to size and focus the image. Position the easel on the enlarger base directly below the projected image.

8. Move the enlarger head up or down until the image is the correct size for the print that will be made. Fine focus the image using a focusing aid, Figure 32-8.

9. Expose a test print. Two common methods are the wedge-strip (often called the test strip) and the print scale.

WEDGE-STRIP PROCEDURE

a. Adjust the aperture to f/11.
b. Place a strip or full sheet of photo paper in the easel.
c. Cover 4/5 of the paper with a piece of opaque cardboard.
d. Give the uncovered strip a 10-second exposure.

e. Move the cardboard so 3/5 of the paper is uncovered.
f. Make another 10-second exposure. Now the first strip will have received 20 seconds of exposure.
g. Repeat the wedge-strip exposures until all five strips of the paper have been exposed. The procedure will give five different exposures of 10, 20, 30, 40 and 50 seconds, Figure 32-9.

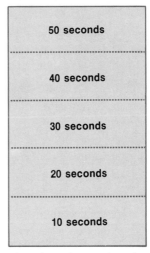

FIGURE 32-9. A wedge-strip test that gives five different exposures.

PRINT SCALE PROCEDURE

a. Adjust the aperture to f/8 or f/11.
b. Place a full sheet of photo paper in the easel.
c. Lay the print scale on top of the photo paper, Figure 32-10. If possible, place it under the frame of the easel to hold it securely in place.
d. Make a 1-minute exposure.

10. Process the test print. Follow the procedure as given in unit 33.

11. Evaluate the test print, Figure 32-11. Decide which wedge strip or pie-shaped segment of the print scale shows the best image. Look for the best highlight and shadow areas. The lightest areas are called *highlights* and the darkest areas are *shadows*. A good photograph contains *tonal values* that approach both ends of a continuous-tone scale (see Figure 24-6).

12. Expose a full print. Set the timer based upon the results of the test print. The minimum exposure time should be 10 seconds. A shorter

FIGURE 32-10. A print scale positioned on top of the photographic paper.

exposure is difficult to control. It is better to reduce the aperture one full stop and keep the exposures over 10 seconds. A recommended exposure range to use is between 20 and 30 seconds.

13. Process the print and evaluate it closely, Figure 32-12. The tonal values may need

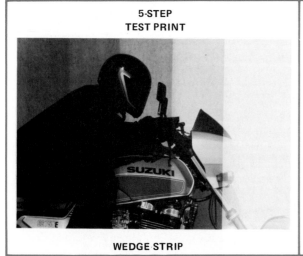

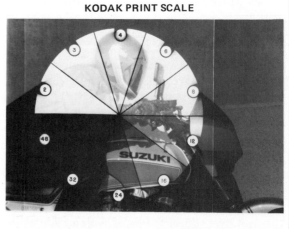

FIGURE 32-11. After processing, test prints must be evaluated to determine the best exposure.

FIGURE 32-12. An enlargement based upon the results of the test print.

improvement; thus, more or less exposure will be needed.

14. Make additional enlargements as needed. Each time a different negative is used, a test print should be made. This will help to conserve on photographic paper.

15. When completed, close down and secure the enlarger. Remember to remove the negative strip from the carrier. Always lower the enlarger head after removing the negative carrier. This helps to keep dust and dirt from getting inside the bellows and on top of the lens.

SUMMARY

Knowing the major parts of an enlarger is important. Also, knowing the types of enlargers, including the several light sources, makes it easier to understand how an enlarger works. Making the first enlargement is an exciting experience. Fol-

lowing procedures that are well thought out give excellent results on any piece of equipment. This is especially true with the enlarger.

REVIEW QUESTIONS

Answer these questions to test your knowledge of the unit content.

1. What is another name for an enlargement?
2. List the four parts that make up an enlarger head.
3. Which type of enlarger is considered the best for making portrait prints?

   A. Combination   C. Condenser
   B. Diffusion   D. Incandescent

4. Which enlarger light source generally has a transformer attached to it?

   A. Incandescent   C. Cold light
   B. Point source   D. High intensity

5. What should be done to the darkroom just before beginning the enlarging procedure?
6. T/F The emulsion of the negative should face down when it is placed into the enlarger.
7. Which test print procedure calls for one exposure?
8. The shortest exposure that should be used to expose a print is:

   A. 30 seconds   C. 50 seconds
   B. 20 seconds   D. 10 seconds

9. A test print should be evaluated by looking at the _____ and _____.
10. T/F Dust and dirt seldom get inside an enlarger, even when the head is left up.

PROCESSING PHOTOGRAPHIC PRINTS

OBJECTIVES   *Upon completion of this unit, you will be able to:*
  • *Describe the processing chemistry used for photographic paper.*
  • *Make the necessary preparations for processing photographic paper.*
  • *Process photographic prints in the conventional manner.*
  • *Produce photographic prints using the stabilization process.*

KEY TERMS   *The following new terms will be defined in this unit:*
Activator Solution
Stabilization
Stabilizer Solution

INTRODUCTION

Processing exposed light-sensitive paper makes the photographs visible. The five stages involved in processing prints are developing, stopping, fixing, washing, and drying, Figure 33-1. Accurate completion of each step gives quality results. Haste and carelessness have no place in the darkroom.

PROCESSING CHEMISTRY

Exposed photographic paper is processed with chemistry in about the same manner as with film. The major difference in the two processing procedures is with the developer. Paper developers tend to be stronger than film developers. This makes them work faster.

Developer solution contains the same four groups of chemicals as does film developer. These

are developing agent, activator, restrainer, and preservative, Figure 33-2. The developing agents commonly used are *metol* and *hydroquinone*. Action of these and other developing agents reduces the exposed silver halide crystals to image-forming metallic silver. The *activator* or accelerator is alkaline, such as sodium carbonate. The purposes of the activator are to speed the developing action and to soften the emulsion slightly. This makes it easier for the developer to penetrate.

The restrainer portion of the developer keeps the developing agent from attacking the unexposed silver halides. Fogging is caused if the restrainer does not perform its task. A typical material used for the restrainer is potassium bromide. Finally, the preservative, such as sodium sulfite, helps to slow the oxidation. Developer must be kept fresh while in the open tray. Oxidized developer will not bring

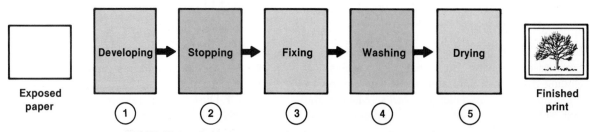

Exposed paper	Developing	Stopping	Fixing	Washing	Drying	Finished print
	①	②	③	④	⑤	

FIGURE 33-1.   The five stages involved in processing a photographic print.

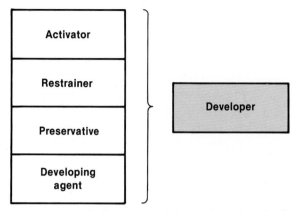

FIGURE 33-2. The four groups of chemicals combined with water to make photographic paper developer.

out image density and can cause surface stains on the print.

Stop bath and fixing solutions are essentially the same as those used for film. The dilutions are often weaker than for film. Specific mixing instructions provided by the manufacturers of the chemistry and paper should be followed closely. Washing aids help to shorten the time needed to remove chemicals from the print. Also, they help to keep the print from curling during the drying stage. Washing aids are used primarily with fiber-based papers.

PREPARING TO PROCESS PHOTOGRAPHIC PRINTS

A critical part of print processing is the preparation. These important steps should be completed before exposing a sheet of photographic paper.

1. Turn on the ventilation fan. It is important to have air circulation in the darkroom.

2. Wear goggles. Obtain and wear a pair of splash-proof goggles. Eye safety is important at all times and especially when working with chemical solutions.

3. Arrange the trays and print tongs, Figure 33-3. Position the four trays in a sink or on a counter. It is best to have them in a sink so that spilled chemistry will run down the drain. The standard arrangement is to process paper and film from left to right. Each tray should be labeled such that the contents can easily be seen under the safelight. One print tong should be placed between each tray. Make certain the trays are large enough for the largest print that will be processed.

4. Mix the processing chemistry. Use a developer recommended by the manufacturer.

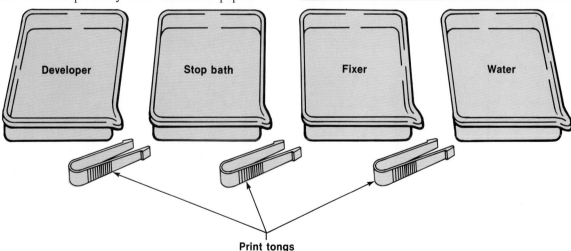

FIGURE 33-3. Processing trays and print tongs arranged in the order of their use, left to right.

FIGURE 33-4. Pouring developer solution into the tray.

Read the instructions very closely on the concentrate packages before mixing the working solutions. Developer concentrates are available in either liquid or powder. Stop bath and fixer strengths must meet the specifications for the photographic paper.

5. Pour in the chemistry, Figure 33-4. Hold the plastic bottle or graduate close to the tray. Pour slowly so that the chemistry does not splatter. Add enough of each chemical solution to make it 1/2 to 3/4 inch (12.5 to 19.0 mm) deep. The actual amount needed for each tray will depend on the tray size.

6. Prepare the washing system. Fasten a washing siphon or another washing unit on the wash tray. Fasten it to a water faucet and start the water into the tray. When the water has reached its operating level, turn the water off. It is a waste of water to keep it running until prints are ready to be washed. Make certain the wash water is between 66 and 70 °F (19 and 21 °C).

7. Establish safelight conditions. Turn on the safelights in the darkroom. Turn off all white lights. Inspect the room for light leaks through the entrance door. Also, take time to inspect all walls and the ceiling for possible white light. Conduct a safelight test. In total darkness, remove a piece of photographic paper from the package. Lay it in a normal working area, such as near an enlarger. Place a coin (nickel or quarter) on the center of the paper. Turn on the safelights for a minimum of 5 minutes. Process the paper. If there is any visible sign of the coin on the paper, the safelights are not safe or there is light coming into the room.

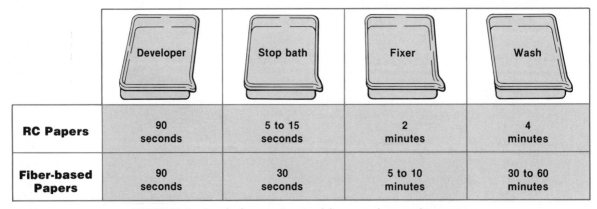

	Developer	Stop bath	Fixer	Wash
RC Papers	90 seconds	5 to 15 seconds	2 minutes	4 minutes
Fiber-based Papers	90 seconds	30 seconds	5 to 10 minutes	30 to 60 minutes

FIGURE 33-5. Standard processing times for most photographic papers.

PROCESSING PRINTS

Standardized processing procedures give consistent results. It is critical that processing times and handling of the prints be the same, Figure 33-5. Controlling processing variables makes it possible to analyze printing results in a more accurate manner. Changes in the total photographic printing procedure should never occur with the processing. The following procedure is for resin-coated (RC) papers.

1. Obtain the exposed piece of photographic paper. It can be exposed by either contact or projection. Be certain to have completely dry hands when picking up the print paper.

2. Develop the print for an exact time. Most papers require 90 seconds, but other times are possible. The minimum developing time is 1 minute. Do not shorten the developing time from that recommended by the manufacturer.

a. With fingertips, hold the paper by the edge.

b. Lift the front side of the tray. This makes the developing chemistry move to the back side.

c. Watch the second hand of a clock. When it is within 5 seconds of the number 12, begin the next step.

d. Push the front edge of the paper under the developer and lower the tray, Figure 33-6. This floods the paper with chemistry and starts the developing action over the entire sheet.

e. Slowly agitate the chemistry by slightly lifting the tray up and down by the ends or sides.

f. Using a print tong, lift the developed print from the chemistry. Do this when 10 seconds of developing time are remaining. Allow the chemistry to drain from the print and back into the developer tray. It is best if print tongs do not touch the image area because they could leave marks in the soft emulsion.

3. Transfer the developed print to the stop bath tray, Figure 33-7. Lay the print in the chemistry. Complete this step as quickly

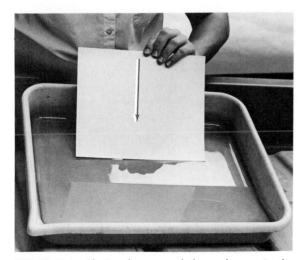

FIGURE 33-6. Placing the exposed sheet of paper in the developer tray.

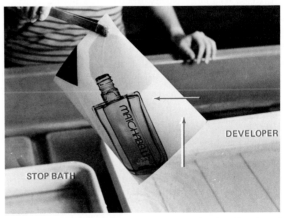

FIGURE 33-7. Using print tongs to transfer the developed print to the stop bath.

as possible. DO NOT allow stop bath solution to get on the tongs. The acetic acid in the stop bath will weaken the alkaline base of the developer when the tongs are used to transfer the next print.

4. Continually agitate the stop bath solution for the full 5 to 10 seconds time. Although the time is very short, this processing step is critical. The stop bath immediately neutralizes the action of the developer.

5. Lift the print from the stop bath with the second print tongs. Allow the solution to drain from the print. Place the print in the fixer tray. Continually agitate the chemistry for the first 30 seconds. Let the print soak for 30 seconds and then agitate for 5 seconds. Repeat this cycle until the 2-minute fixing time has elapsed.

6. Using the third print tongs, remove the print from the fixer and place it in the wash tray for 4 minutes. Watch the time closely as a longer washing period may be damaging to the photographic paper. Too much washing will cause the emulsion and paper base to separate. It is advisable to use a holding tray filled with water for fiber-based paper. With the longer wash cycle, it is unwise to place freshly fixed prints in the wash tray until the batch is completed and removed.

7. Dry the photographic print.
   a. Remove the print from the wash bath. At this point, it is appropriate to use fingers instead of print tongs. Allow excess water to drain off.
   b. Place the print on a hard smooth surface of glass, plastic, or stainless steel. For best results, the surface should be higher at one end so water will drain from the print.

FIGURE 33-8.   Removing surface water from a print with a soft-blade squeegee. *Doran Enterprises, Inc.*

   c. Carefully remove water from the print using a soft-blade squeegee or a chamois, Figure 33-8.
   d. Place the print on a dry rack within the RC print dryer, Figure 33-9. Close the door and turn on the heater and fan. Low heated air gently moving over the print will dry prints in 30 to 60 minutes.

8. Another method of drying RC prints is shown in Figure 33-10. With this system, the wet print is rolled between two squeegee-type rollers. This removes the excess water. The prints are then placed in the drying racks which are fastened to the wall.

9. Fiber-based prints are dried on a heated ferrotype plate type of dryer.

10. Clean up the darkroom when all prints have been processed for the given work period.

FIGURE 33-9. Placing a print in an RC dryer.

CLEAN-UP PROCEDURE

a. Pour the developer and stop bath solutions into a holding tank for proper disposal at a later date.

b. Pour the fixer into the "used" fixer container for future use. When the fixer is exhausted, it too should be poured into the holding tank.

c. Empty the water wash tray contents down the drain.

d. Wash and dry the trays. Also, return them to their storage places.

e. Clean the working counter, sink, and entire darkroom.

f. Return safety goggles to their proper storage location.

g. Turn off all lights and the ventilation fan.

PRINTS BY STABILIZATION

The *stabilization* process makes it possible to produce photographic prints in a very short period of time. This is possible because of the special photographic paper. Developing agent, often hydroqui-

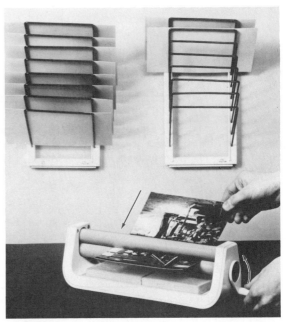

FIGURE 33-10. Drying RC prints using a squeegee roller and rack system. *Falcon Safety Products Inc.*

none, is included directly into the paper emulsion. Development takes place when the exposed paper is immersed into a highly alkaline *activator solution.* After development, the prints are fixed in a *stabilizer solution* that neutralizes the activator.

Stabilization prints are always processed with specially designed equipment, Figure 33-11. The stabilization processor contains two bath trays

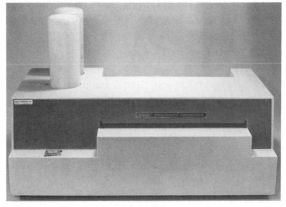

FIGURE 33-11. A two-bath stabilization print processor.

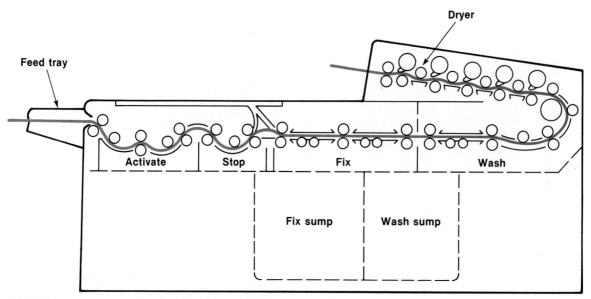

FIGURE 33-12. A schematic diagram of a stabilization processor that is used to produce permanent prints in 55 seconds. *Eastman Kodak Company*

that hold the activator and stabilizer solutions. The processor also contains several rollers that carry prints through both solutions. Prints are processed in about 15 seconds, but they have a short life. The life of stabilization prints can be increased similar to conventional processed prints by regular fixing, washing, and drying.

Complete rapid processing of stabilization material is done by using resin-coated paper and a larger processor. The paper contains the same basic layers as standard resin-coated paper. The major difference is that the emulsion contains the developing agent. After exposure in the normal manner, prints are placed in a processor that completes all five processing steps, Figure 33-12. The total processing time is just less than 1 minute.

SUMMARY

Photographic paper processing chemistry and film processing chemistry are much the same. The major difference is the strength of each. Before paper processing can begin, it is important to make complete preparations. Quality processing is only possible when each and every step is done with care. Rapid processing is done through a process called stabilization. Finished products (prints) can either be made in the conventional manner or through high-speed stabilization. The results look the same; thus, a choice of two processes is available.

REVIEW QUESTIONS

Answer these questions to test your knowledge of the unit content.

1. Name the five stages of processing photographic prints.
2. What is the major difference between paper and film developer?
3. What problem can occur if the restrainer portion of the developer solution does not work?

   A. Under development   C. Fogging
   B. Oxidation   D. Bleaching

4. T/F Washing aids should be used when processing resin-coated papers.

5. T/F Processing trays and chemistry should be arranged so the work flow moves from left to right.

6. The temperature of the wash water for prints should be between _____ and _____ °F or _____ and _____ °C.

7. Explain how to conduct a safelight test in a print processing darkroom.

8. The standard developing time for most photographic papers is:

A. 30 seconds   C. 90 seconds
B. 60 seconds   D. 120 seconds

9. _____ _____ should be used to transfer prints from one processing tray to another.

10. What is special about stabilization paper that makes it develop so rapidly?

UNIT 34

CONTACT PRINTING AND PROOFING

OBJECTIVES   *Upon completion of this unit, you will be able to:*
- *Describe the four contact printing methods.*
- *Select a light source and make a contact proof.*
- *Analyze a contact proof sheet in preparation for making enlargements.*
- *Prepare a creative photogram.*

KEY TERMS   *The following new terms will be defined in this unit:*

Contact Print   Printing Frame
Contact Printer   Photogram
Print Proofer

INTRODUCTION

Contact printing is a simple process. It is done by placing a negative against a piece of photographic paper and exposing the paper to light, Figure 34-1. Once the paper is exposed, it is processed to make the positive image visible. *Contact prints* are always the same size as the negatives. This makes them most useful for proofing purposes. This is especially true for small size film such as 35 mm.

CONTACT PRINTING METHODS

Four basic methods can be used to make contact prints. Each of these methods gives excellent results. The main differences involve convenience and cost.

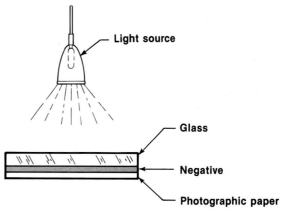

FIGURE 34-1. The contact printing arrangement.

FIGURE 34-3.   A printing frame used to press negatives and photographic paper tightly together. *Doran Enterprises, Inc.*

FIGURE 34-2.   A contact printer that is completely self-contained. *Omega Arkay*

CONTACT PRINTER.   This is a self-contained exposure unit, Figure 34-2. It has a built-in light source, a ground glass, a pressure pad, and a timer. Exposures can be made very rapidly with a unit of this type.

PRINTING FRAME.   A frame, generally made of wood, a piece of glass, and a pressure plate make up a *printing frame*, Figure 34-3. Negatives and photographic paper are held tightly together between the glass and pressure plate. The exposure is then made using a separate light source.

PRINT PROOFER.   A hinged glass frame and a pressure pad base are the main parts to a unit of this design, Figure 34-4. Photographic paper is placed on the base with the emulsion side up. Negatives are placed over the paper with the emulsion side down. Some print proofer glass frames have channels for the negatives. This helps to align the negatives and give them equal spacing. The frame

FIGURE 34-4.   A print proofer (contact printing frame) designed to handle 35-mm negatives. *Porter's Camera Store, Inc.*

is closed, and the exposure is made using a separate light source.

PLATE GLASS.   A piece of plate glass works well when making contact proofs. For safety purposes, it is important that the edges and corners be smooth. The glass should be at least 2 inches (5 cm) larger than the paper on each of the four sides.

LIGHT EXPOSING SOURCES

Three types of lights are commonly used to make contact exposures, Figure 34-5. The white wall or

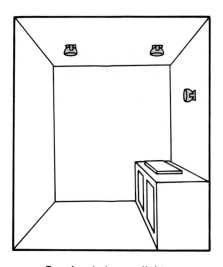

Regular darkroom lights

Bulb and reflector

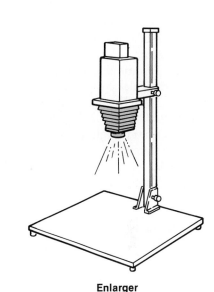

Enlarger

FIGURE 34-5.   Common light sources for exposing contact prints.

ceiling lights in a darkroom are the least effective. Location and intensity make them somewhat unreliable. The bulb with a reflector is a good light source. Exposure control can be obtained through bulb size, the distance that the bulb is above the photographic paper, and length of exposure.

The most effective exposure light source is the enlarger. Complete control of the exposure is possible. Aperture adjustments permit intensity control. The timer attached to the enlarger gives accurate exposure length. The height of the enlarger head is easily adjusted. This permits full control of the light coverage. Placing a film holder in the enlarger without a negative gives the projected light a rectangular shape. This keeps light from spreading to other areas of the darkroom and fogging other photographic materials.

PROCEDURE FOR PRODUCING CONTACT PRINTS

Several steps are necessary to produce quality contact prints. The following steps lead to results that can be very useful to a photographer. An enlarger is used as the light source.

1. Obtain needed materials. These include the negatives that have been protected in plastic or paper negative sleeves and photographic paper. A good paper to use is resin-coated, variable-contrast or number 2, glossy finish, and 8″ × 10″ (20.3 × 25.4 cm) in size. A piece of plate glass can be used to hold the negatives and paper in contact.

2. Prepare the processing chemistry as presented in unit 33. Wash and dry hands thoroughly after working with chemical solutions. Never work with negatives and photographic paper with wet hands.

3. Adjust the enlarger head height so the lens is 20 to 24 inches (50.8 × 61 cm) from the enlarger base.

4. Set the lens aperture at f/11.

5. Insert a 35-mm negative carrier in the enlarger head. No film should be in the negative carrier.

6. Establish safelight conditions in the darkroom.

7. Turn on the enlarger light and focus the rectangular area of light on the base, Figure 34-6. The light should cover an area about 10″ × 12″ (25.4 × 30.5 cm). Adjust the enlarger head height if necessary. This area must be larger than the paper that will be exposed.

8. Place pieces of masking tape at the two far corners of the rectangular lighted area, Figure 34-7. The tape serves to identify the size of the lighted area. It helps in locating the photographic paper, negatives, and glass when making an exposure.

9. Turn off the enlarger light, and adjust the switches on the timer for 5 seconds.

10. Lay a piece of 8″ × 10″ (20.3 × 25.4 cm) photographic paper on the enlarger base. Position it within the area that will be lighted by the enlarger. Make use of the two corners marked with masking tape. Be certain that the emulsion of the paper is up.

11. Position negative strips (emulsion down) on the photographic paper, Figure 34-8. Remove the negatives from the protective

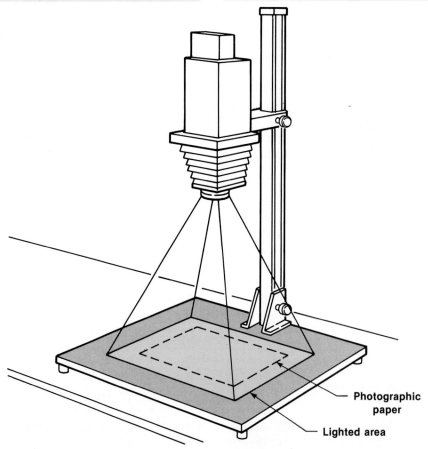

FIGURE 34-6. The lighted area on the base must be larger than the photographic paper.

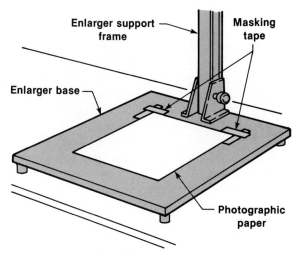

FIGURE 34-7.   Masking tape at the two far corners of the rectangular lighted area serve as guides to position the paper.

FIGURE 34-8.   The negative strips are placed on top of the photographic paper for contact printing.

sleeve. It is possible to make contact exposures with negatives in the protective sleeves, but the quality of the contact prints will be reduced. The extra thickness of material between the negative and photographic paper allows the light to spread. Best results in contact printing are obtained when the materials are placed "emulsion-to-emulsion."

12. Lay the piece of glass over the negative strips and photographic paper. This must be done with care so that the negative strips do not move. Make certain the negative strips and paper are in tight contact.

13. Make a test exposure. A typical wedge test gives three different exposure lengths of 10, 15, and 20 seconds (see explanation of procedure in unit 32).

14. Lift the glass and stand it on edge against the vertical frame of the enlarger. Also, remove negative strips from the photographic paper. Place them on the outer edges of the enlarger base in readiness for use again.

15. Process the test print. Follow the procedure as given in unit 33.

16. Inspect the test print and determine the best exposure, Figure 34-9. Select a time that

FIGURE 34-9.   The contact test print should be inspected to determine the best exposure time.

gives the best highlight and shadow tonal values.

17. Expose and process a proof sheet using the best exposure time. Exposing several negative strips at one one time on a piece of photographic paper saves time both in exposing and processing.

18. When finished, clean and secure the darkroom.

ANALYZING THE PROOF SHEET

When the proof sheet is completely dry, it is time to closely inspect each of the negative frames. The following steps help to organize this important phase of contact proofing.

1. Look at the entire proof sheet. Determine if one or more of the pictures jump out and say, "print me." If so, make an X on the frame(s) with a felt-tip marker.

2. Use a paper mask to view each proof frame, Figure 34-10. The mask blocks out all of the other proofing frames. This gives a clear view of only one frame at a time. Make a paper mask by cutting a rectangular, 35-mm size opening in the center of a piece of typing paper.

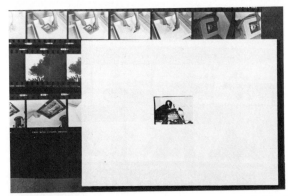

FIGURE 34-10. A mask is useful for viewing individual frames on a contact proof sheet.

3. Select specific frames for enlarging. There may be several frames that have good content. There is no set number of proofing frames that must be selected. It is best to be critical and select a few good frames than to select a large number of medium quality frames.

4. Mark each selected frame with an "X." Use a black felt-tip marker and place the X in a light area of the selected frame. This makes it easier to see under the safelight in the darkroom.

5. Trim excess photographic paper from the proof sheet. Use clear tape and fasten the proof to the envelope sleeves used for the negative strips that were proofed. This keeps the proofs and negatives together for later use and reference.

MAKING PHOTOGRAMS

Photograms are photographic prints made without using negatives, Figure 34-11. They are made by placing objects on photographic paper and exposing the paper to light. The procedure is listed as follows.

1. Obtain several small objects, such as coins, paper clips, keys, jewelry, etc.

2. Prepare the enlarger to make the exposure. Place a film holder, without film, in the enlarger. A 35-mm film holder works very well. Adjust the height of the enlarger head so that the rectangular lighted area covers the selected paper size.

3. Place a piece of photographic paper in the easel.

4. Position several objects directly on the emulsion of the photographic paper, Figure 34-12. Be very careful to not touch or scratch the paper emulsion. Place the objects in a creative arrangement.

FIGURE 34-11. Photograms are made without using a negative but can be very creative.

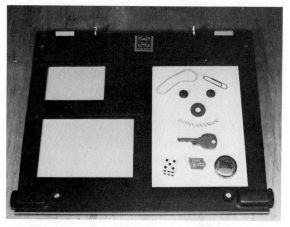

FIGURE 34-12. The opaque and semi-opaque objects should be carefully positioned on the photographic paper to make a creative photogram.

5. Make the exposure. The length of time will vary depending on the f-stop and the height of the enlarger head. A good starting point is f/11 for 20 seconds.

6. Remove the objects from the photographic paper. Do this very carefully.

7. Process the exposed print in the usual manner.

8. Evaluate the results. The exposed emulsion should be a dense black. If the emulsion is light, more exposure is needed. Less exposure is needed if light penetrated some of the objects used to create the image.

SUMMARY

Contact printing and proofing are an important part of print making. The contact prints are always the same size as the negatives. This makes 35-mm negative prints small, but they serve very well as proofs. After studying the proofs, selected frames can be made into enlargements with confidence. This procedure saves much time and material cost. Making photograms is a form of contact printing that gives good results and satisfaction.

REVIEW QUESTIONS

Answer these questions to test your knowledge of the unit content.

1. Why is a contact proof the same size as the negative?

2. Which contact printing method is most likely the least expensive?

    A. Plate glass   C. Print proofer

    B. Contact printer   D. Printing frame

3. Contact proofs are made with the negative and paper emulsions facing:

   A. Away from each other
   B. Up or down
   C. The light source
   D. Each other

4. Which exposure light source provides the greatest amount of control?

5. T/F Contact prints are developed and processed in the same manner as projection prints.

6. What item should be placed in the enlarger to help control the light when making contact exposures?

7. Which f-stop is a good starting point when making contact exposures with an enlarger?

   A. f/32   C. f/5.6
   B. f/11   D. f/2

8. Why must the negatives and print paper be held in tight contact when making an exposure?

9. A _____ helps to analyze contact proof prints.

10. T/F Photograms are often made by using small transparent objects.

— UNIT 35 —

IMPROVING PHOTOGRAPHIC PRINTS

OBJECTIVES   *Upon completion of this unit, you will be able to:*
- *Use the five methods of image control to improve photographic prints.*
- *Determine the best overall contrast control methods to use for given situations.*
- *Prepare finished photographic prints using localized contrast control methods.*
- *Analyze photographic prints and determine the problem cause and solution for each.*

KEY TERMS   *The following new terms will be defined in this unit:*

Bleaching   Distortion   Printing-in
Cropping   Dodging   Vignetting
Diffusion

INTRODUCTION

Making a basic photographic enlargement is an enjoyable experience. Improving the basic print adds more excitement and satisfaction. Learning and using the several "improving" methods and techniques are the key to success in black-and-white printmaking.

IMAGE CONTROL

The appearance of photographic enlargements can be greatly improved by using one or more methods of image control. The typical methods of controlling images are cropping, distortion, diffusion, and vignetting.

   CROPPING.   This involves the removal or nonprinting of unwanted and distracting elements within the negative. An enlargement of the entire negative may include content that distracts from the focal point of the image, Figure 35-1. Raising the enlarger head and refocusing the image removes the unwanted image area, Figure 35-2. This also increases the size of the desired image area.

   Printed photographs in books, newspapers, and magazines often are cropped. Just as with cropping in the enlarger, the main purpose is to remove distracting elements of the image. Often, cropping for publication serves to make the photograph fit a given image area.

A pair of L-shaped paper or thin plastic masks help to select the desired content for a photograph, Figure 35-3. The masks can easily be moved from side-to-side and up-and-down until the best appearing image is obtained. Short marks called "crop marks" are then placed in the print margins. These small marks are used as guides when making a new enlargement. They also can be used by graphic arts printers when preparing the photograph for printing in a publication. L-shaped masks also can be very useful in determining cropping directly on a contact proof sheet.

DISTORTION. This method of image control is used when image lines tend to converge (get closer together). Taking pictures of tall buildings with a 35-mm camera often results in the top looking smaller. Also, the top edges of the building move closer together (see Figure 18-13).

Distortion problems often can be corrected while exposing the print with the enlarger.

1. Prepare the enlarger in the usual manner.

2. Size, crop, and focus the image on the easel.

3. Raise the side of the easel on which the image width is the greatest. Note that the image becomes smaller as it gets closer to the enlarger lens.

4. Find the best height for the raised side of the easel. Place a wood block under the easel so it will stay in position, Figure 35-4.

FIGURE 35-2. Cropping distracting elements from the photograph shown in Figure 35-1 improves the image and makes a stronger picture.

FIGURE 35-1. A print of the entire negative may contain distracting elements.

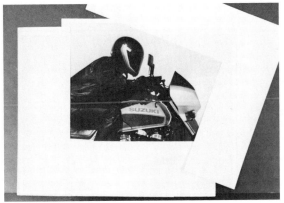

FIGURE 35-3. L-shaped masks help when cropping a photograph.

5. Refocus the image. Select an area about 1/3 distance down from the raised side as the focusing point.

6. Stop the lens down to the smallest f-stop possible. This will give the greatest depth of field.

7. Expose and process a print. It will be necessary to conduct test exposures as with any print. Also, the exposure technique known as "dodging" must be used. This is explained later in this unit. Dodging is required because the raised side will require less exposure than the lowered side.

DIFFUSION. Reducing the sharp, harsh tones of an image is the purpose of the diffusion technique. One of the easiest ways is to use a special print effect filter in the enlarger, Figure 35-5. The filter is placed in the filter holder directly under the lens. It tends to soften the image edges and the harsh lines within the image.

Other diffusing methods include fabric netting, thin plastic sheets, and glass smeared with translucent jelly. Nylon netting such as that used for stockings can be stretched tight over a wire frame.

1. Set up the enlarger in the usual manner.

2. Focus, size, and crop the image on the enlarger.

3. Place a sheet of print paper in the easel.

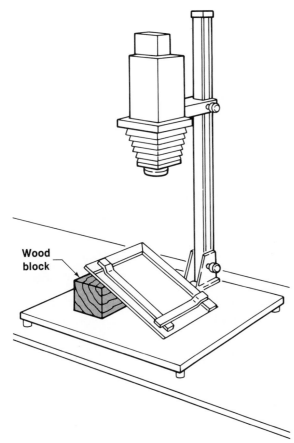

Wood block

FIGURE 35-4. A side of the easel is raised to help correct image distortion contained in the negative.

WITHOUT PRINT EFFECT FILTER

WITH PRINT EFFECT FILTER

FIGURE 35-5. Using a print effect filter in an enlarger softens the harsh image edges. *Tiffen Manufacturing Corp.*

4. Hold the wire frame with the stretched nylon about half way between the enlarger lens and the easel, Figure 35-6. The same can be done with a thin plastic sheet and with smeared glass.

5. Expose the image to the paper. Keep the nylon netting moving in a tight circular motion during the time of this exposure. The movement keeps hard image lines from forming.

6. Process the print in the usual manner.

VIGNETTING. A mask, either commercially purchased or self-made, is used to make a vignetted print, Figure 35-7. A vignetting mask is generally made in an oval shape. It includes serrated edges that help to keep the image edges from being hard lines. Follow the same procedure as with diffusion to make a vignetted print. Moving the mask slowly up and down for a short distance gives soft, smooth image edges.

SPECIAL MASKS. Masks of many shapes and sizes can be handmade or purchased from commercial sources, Figure 35-8. These and many other interesting shapes give added dimension to large and small photographic prints.

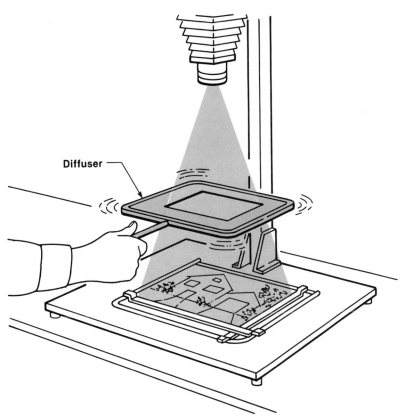

FIGURE 35-6. The diffuser should be held about halfway between the lens and the easel. The diffuser must be kept moving in a tight circle during exposure.

FIGURE 35-7. Vignetting causes the image to stand out strong on the photographic print.

FIGURE 35-8. Masks of various shapes and sizes add much to finished prints.

OVERALL CONTRAST CONTROL

Contrast is important to all photographers. In black-and-white photography, contrast between the many image parts makes it possible to see the image content. Three overall contrast control methods are in common use.

PAPER GRADES. The emulsions of graded papers are formulated to produce prints with a consistent range of contrasts between the highlights and the shadows. The numbered papers permit convenient selection. A number 2 paper gives average contrast and should be used when printing with an average contrast negative. When the negative has been overdeveloped and contains considerable contrast, it is useful to use a paper grade less than 2. For flat, limited contrast negatives, paper contrast

grades over 2 are available. The common range of paper grades ranges from 1 through 5. Some papers are identified with word descriptors—soft (1), medium (2), hard (3), extra hard (4), and ultra hard (5).

USING VARIABLE-CONTRAST PAPERS AND FILTERS

These papers provide convenient contrast control when used with compatible filters, Figure 35-9. Use them as follows:

1. Set up the enlarger in the usual manner.
2. Focus, size, and crop the selected image.
3. Place a number 2 filter in the filter holder below the enlarger lens, Figure 35-10.
4. Make the needed test exposures to arrive at the best exposure time.

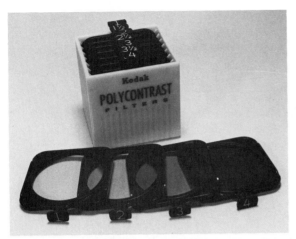

FIGURE 35-9. Variable-contrast filters used with variable-contrast print paper make it convenient to control image contrast.

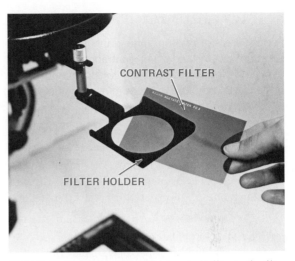

FIGURE 35-10. Placing a variable-contrast filter in the filter holder directly below the lens.

5. Expose and process a print using variable-contrast paper and the number 2 filter. If there is need to increase the contrast, decide how much. A typical set of contrast filters are available in half-steps from 0 to 5.

6. Select a higher number filter than 2 to increase the contrast and a lower number to decrease.

7. Calculate the new exposure length based upon the standard number 2 filter, Figure 35-11.

8. Expose and process the new print. Evaluate and correct as desired.

BLEACHING A DEVELOPED PRINT

A flat, low-contrast print may be bleached to increase its contrast. The bleach will remove density from the highlight areas. This extends the range of tones to give a higher contrast print.

1. Prepare a very dilute solution of potassium ferricyanide. Pour the solution in a standard print processing tray.

2. Expose and process the print in the standard manner.

Filter Number	Exposure Percentage	Example Seconds
1	107	$21\frac{1}{2}$
$1\frac{1}{2}$	107	$21\frac{1}{2}$
2	100	20
$2\frac{1}{2}$	112	$22\frac{1}{2}$
3	145	29
$3\frac{1}{2}$	180	36
4	370	74

FIGURE 35-11. Exposure times for other variable-contrast filters can be determined when the time is known for the number 2 filter.

3. Under white light, immerse the print in the bleach solution. Agitate the solution slightly.

4. Stop the bleaching action by returning the print to the fixer.

5. Wash and dry the print in the usual manner.

LOCALIZED CONTRAST CONTROL

Specific areas of photographic prints sometimes can be improved by adding or subtracting image content. Knowing how to use the four common methods makes it possible to produce quality prints even from questionable negatives.

DODGING

This is a subtractive method of improving a print. It involves keeping the enlarger light from reaching one or more areas of the print paper during exposure.

1. Make an enlargement in the standard manner. Use graded paper or variable-contrast paper and filters to produce a print that is as acceptable as possible.

2. Study the print. With a dark felt-tip marker, circle the area(s) that lack detail because of too much exposure.

3. Plan to make a second print. Use an exposure that is at least 30 seconds long. A long exposure is needed so that there will be time to dodge (subtract) some of the image light. Typical dodging may take between 5 to 15 seconds.

4. Obtain or make a dodging tool. A circle or other shape cut from black stiff paper and fastened to a wire works very well.

5. Place a piece of paper in the easel and begin the exposure.

6. Use the dodging tool and block out 1/3 of the exposure light in the selected areas, Figure 35-12. The dodging tool should be positioned about halfway between the lens and the easel. Keep it moving in a tight circular motion for the desired dodging time.

7. Process the print and study the results. Make adjustments as needed.

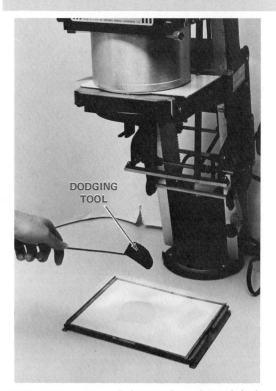

DODGING TOOL

FIGURE 35-12. Using a dodging tool to subtract light from selected image areas.

PRINTING-IN. This method is exactly the reverse of dodging. *Printing-in*, sometimes called burning-in, is an additive method because image light is added to selected areas of a print. Instead of using a dodging tool to subtract light, a mask is used to add light, Figure 35-13. Commercial masks are available, but they can easily be made from stiff black paper. Simply cut a round or irregular shaped hole in the paper. Use the mask and follow the same procedure as outlined for dodging. Remember to keep the mask moving at all times so hard image lines are avoided.

BLEACHING SMALL AREAS. A weak solution of potassium ferricyanide can be used to bleach small areas of a print, Figure 35-14. Just as with bleaching the entire print, small areas are treated after the fixing stage. Use a small brush and apply the bleach sparingly. Stop the bleaching action by returning the print to the fixer tray.

FIGURE 35-13. A commercial easel and localized contrast control device. Note that the printing-in mask conforms to the image created by the bride. *Jobo Fototechnic, Inc.*

FIGURE 35-14. Small areas of a print can be bleached with a dilute solution of potassium ferricyanide.

CONCENTRATED DEVELOPMENT. Heat makes developer chemistry work faster. Using a finger to rub a selected area of a print while it is in the developer solution will give more density to the selected area, Figure 35-15. Body temperature is about 30 °F (16 °C) warmer than the chemistry and paper. This makes a significant difference in the developing temperature of the area being rubbed.

ANALYZING PRINTS

The information found in Figure 35-16 should help solve printmaking problems. Identifying problems and determining causes lead to successful solutions.

SUMMARY

Several methods are available for improving photographic prints. These include five methods of image control, three methods of overall contrast control, and four methods of localized contrast control. Se-

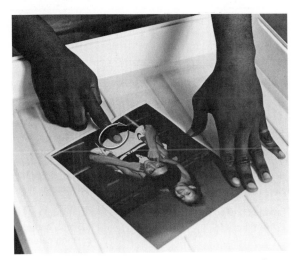

FIGURE 35-15. Rubbing selected areas of a print during development increases the density due to the added heat.

ANALYZING PRINTS		
Problem	**Cause**	**Solution**
Image out of focus	Enlarger lens not focused prior to exposure	Take time to focus and use a focusing aid (magnifier).
Double image	Vibration of enlarger head during exposure	Place enlarger on a solid table or counter. Also, do not touch enlarger during exposure.
White specks within image	Dust on film and/or on glass negative carrier	Vacuum darkroom and enlarger on a regular basis. Handle film with care and keep it clean.
Highlights and border gray	Safelights or other stray light fogging print paper	Test safelights to determine their light-safe condition. Cover other light leaks to darkroom.
Dark streaks and areas	Stray light from the enlarger during exposure reaching the print paper	Cover light leaks from enlarger head with heat resistant black tape. Surround enlarger on sides and back wall with black paper or cloth to absorb stray light.
Overall image light with little density in shadow areas	Exposure too short or developer weak	Increase exposure length and/or change to new developer. Note: Change only one variable at a time so the cause can be isolated.
Overall image too dark	Exposure too long or developer too strong	Decrease exposure length and/or change developer.
Limited contrast between highlights and shadows	Negative underexposed and/or underdeveloped	Use a higher contrast graded print paper or a higher contrast filter with variable-contrast paper.
Selected area of print too dark	Negative contains thin area due to uneven exposure	Dodge area during exposure or spot bleach after fixing.
Selected area of print too light	Negative contains dark area due to uneven exposure	Print-in area during exposure or rub area during development to increase developing action.
Distracting image elements within the picture	Distracting elements unnoticed during picture taking or unable to remove distractions due to conditions	Crop the image on the negative while making the print on the enlarger.
Image lines on a tall object—building or tower—tend to converge	Picture taken with a small format (35 mm) camera. Also, taken too close causing camera lens to be pointed upward	Use distortion control to correct image.

FIGURE 35-16.   This chart is useful in determining solutions for problems that occur in printmaking.

Stains on print surface	Oxidized developer or over-development of print	Maintain developer strength. Change to new developer after recommended maximum number of prints have been processed.
Prints turn yellow and dark after processing has been completed	Exhausted fixer	Maintain fixer strength. Change to new fixer when the old batch appears dark.

FIGURE 35-16. (Continued)

lecting one or more of the specific methods depends upon the conditions and needs for the finished prints. Analyzing completed prints is valuable for improving the finished product—the photograph.

REVIEW QUESTIONS

Answer these questions to test your knowledge of the unit content.

1. Which image control method involves the removal of distracting image content?

   A. Cropping   C. Diffusion
   B. Distortion   D. Vignetting

2. T/F Cropping is always done with an enlarger.
3. Which image control method involves the raising of an edge of the easel?

   A. Cropping   C. Diffusion
   B. Distortion   D. Vignetting

4. T/F A vignetting mask should always be kept moving during an exposure.

5. Which paper grade gives the highest contrast?

   A. 1   C. 3
   B. 2   D. 4

6. Which variable-contrast filter provides the average print contrast?

   A. 1   C. 3
   B. 2   D. 4

7. A number 4 variable contrast filter requires—more or less—exposure as compared to a number $2\frac{1}{2}$ filter.
8. A photographic print is bleached after the print is _____.
9. Which of the localized contrast control methods subtracts light?

   A. Printing-in
   B. Bleaching
   C. Dodging
   D. Concentrated development

10. What is the first step in analyzing a photographic print for the purpose of improvement?

CHAPTER SEVEN

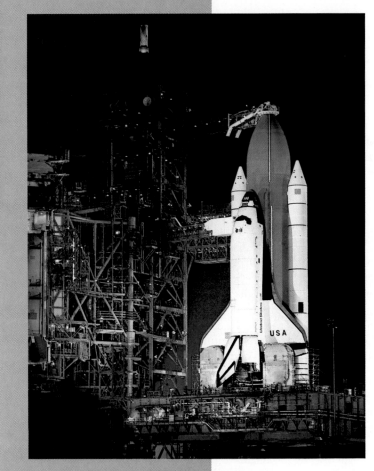

COLOR PHOTOGRAPHY

TAKING COLOR PHOTOGRAPHS

OBJECTIVES   *Upon completion of this unit, you will be able to:*
- *Explain the basic principles of color.*
- *Describe the two general categories of color film.*
- *Select special equipment suitable for taking slides.*
- *Take quality pictures in full color.*

KEY TERMS   *The following new terms will be defined in this unit:*

Additive Colors	Copy Stand	Positive Color Film
Color Wheel	Macrophotography	Subtractive Colors
Complementary Color	Negative Color Film	

INTRODUCTION

Color is everywhere. It is present in nature throughout land, sea, and sky. People-made products also include much color in textiles, plastics, paints, and photographic materials. Color film and photographic paper make it possible to record all surrounding color in a very accurate manner, Figure 36-1. The first "easy to use" color films became available in the mid-1930s. Eastman Kodak Company released their *Kodachrome* (slide film) in 1935 and Agfa-Gevaert, Inc. released their *Agfacolor* (print film) in 1936. These and other films that appeared following World War II gave people every opportunity to capture the world in color.

BASIC PRINCIPLES OF COLOR

Seeing color materials is possible because the eyes and brain working together can separate them. Lighting has a critical effect on how color is seen. Sunlight is the best light to view all colors found in the visible spectrum. It serves as the standard by which all artificial light is measured. Sunlight varies in intensity and color depending on the time of day. Nothing can be done about controlling sunlight of course, but it is possible to make adjustments to meet the requirements of color film.

The intensity of sunlight varies depending on the time of day, cloud cover, and season of the year. Light meters make it convenient to measure

FIGURE 36-1.   Color photographic film and paper make it possible to record colors accurately even under low light conditions that are found in our environment. *Julie Habel*

FIGURE 36-2. Late afternoon color photographs often contain the warm colors of sunset.

FIGURE 36-3. The three primary colors of light are called additive colors.

sunlight intensity. This generally causes the photographer few problems unless care is not taken to take accurate light meter readings. Refer to unit 16, "Measuring Light" to review this important aspect of both black-and-white and color photography.

The color of light is measured on the Kelvin scale (see Figure 19-10). Film is color-balanced for the average color temperature (5500 °K) of sunlight which occurs during the middle parts of the day. Early morning and evening sunlight tend to contain high levels of yellow, Figure 36-2. A hazy and overcast sky tends to be bluish in color. This does have an effect on how color film records colors. Decisions when to take a color picture should be based upon the "color" wishes of the photographer. Many times, unfortunately, the luxury of choosing the sunlight conditions is not possible.

ADDITIVE COLORS.   The human eye is sensitive to the three primary light colors: red, green, and blue, Figure 36-3. These three colors have been created from a blend of the spectrum in equal 1/3 amounts. Mixing the three colors in equal amounts creates white, Figure 36-4. This is the basis for the statement that "white is the presence of all color." Combining red and green in equal amounts creates the secondary color of **yellow**. Combining green

and blue in equal amounts creates the secondary color of **cyan**. Finally, combining blue and red in equal amounts creates the secondary color of **magenta**. The term *additive* then describes colors that add color instead of taking anything away. Color television is based upon additive color theory.

SUBTRACTIVE COLORS.   The three main *subtractive colors* are yellow, cyan, and magenta, Figure 36-5. These are created from the primary additive colors of red, green, and blue. Combining two subtractive colors recreates one of the primary additive colors. Combining all three of the subtractive colors in equal amounts gives a near black, Figure 36-6. This is the basis for the statement that "black is the absence of all color." Subtractive colors do actually subtract some color, but they do allow the majority of color to pass through. Color film is based upon subtractive color theory.

A *color wheel* helps to explain the relationship between the additive and subtractive colors, Figure 36-7. The primary light (additive) colors form a triad. A subtractive color is positioned between each two additive colors. This too forms a triad of the subtractive colors. The following statements can be said, as well as seen, for each of the subtractive colors:

- Yellow = red and green but minus blue
- Cyan = green and blue but minus red
- Magenta = blue and red but minus green

Each primary color in the additive and subtractive color wheel has a complement. That color

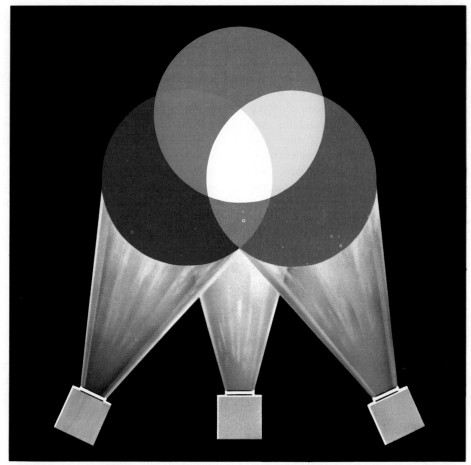

FIGURE 36-4. White light results when the three primary light colors are combined in equal amounts. The three secondary colors—yellow, cyan, and magenta—result when two primary colors are combined. (Reprinted from *Karsnitz, Graphic Arts Technology,* Figure 17-2.)

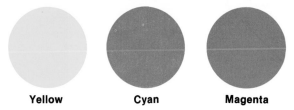

| **Yellow** | **Cyan** | **Magenta** |

FIGURE 36-5. The three main subtractive colors.

is found directly opposite any given color. For example, the complement of green is magenta, which is formed by combining the other two primary colors. The complement of red is cyan, and the complement of blue is yellow. The subtractive colors are often referred to as *complementary colors.*

COLOR FILMS

A wide variety of color films is available to amateur and professional photographers. This gives everyone having a camera sufficient opportunity to select the film of their choice. Color films can be easily divided into two groupings according to their use.

POSITIVE COLOR FILM. This type of film yields right reading images in full color. These films are most often referred to as slide films because the processed film serves as the finished image. The individual exposures are mounted into 2″ × 2″ (5 × 5 cm) cardboard or plastic frames for use in viewing on a screen with a slide projector. Positive color films also are referred to as reversal films. Another term often used is "transparency" film. The reversal term primarily stems from the type of chemical processing used for positive color films. Film names often include the term "chrome" to designate that it is to be used for slides. Kodachrome, Agfachrome, and Fujichrome are typical examples. Some companies clearly label the film box and magazine as "color slide" so that no errors will be made by the user.

NEGATIVE COLOR FILM. These films are most often designated by using "color" as part of the name. Such film names as Kodacolor, Agfacolor, and Fujicolor inform the user that the film will give negatives from which color prints can be made.

All color film contains three or more emulsion layers, Figure 36-8. Each of the layers is sensitive to one or more of the primary additive colors. The top yellow layer is sensitive to blue light. The layer immediately below the blue-sensitive emulsion serves as a filter to absorb excess blue

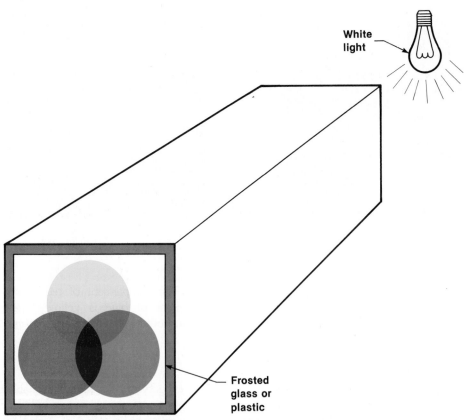

FIGURE 36-6. A near black image results when the three main subtractive colors are combined in equal amounts. A primary additive color is recreated when two subtractive colors are equally combined.

light. This is necessary because the green-sensitive emulsion (magenta) and the red-sensitive emulsion (cyan) are also sensitive to blue light. Several color films contain multiple color-emulsion layers. This permits greater accuracy in recording the true image colors during exposure.

CARE OF COLOR FILM. The expiration date printed on film packages has meaning. It means that under proper storage, the film will faithfully reproduce the intended colors. Proper storage means low humidity and a cool temperature. It is always best to keep color film in an area that has less than 50% relative humidity. Also, the temperature should range between 50 °F (10 °C) and 68 °F (20 °C). Heat and humidity are damaging to film.

SPECIAL EQUIPMENT FOR SLIDES

A wide variety of equipment is available for color photography. This is especially true for making slides. The popularity of preparing and using color

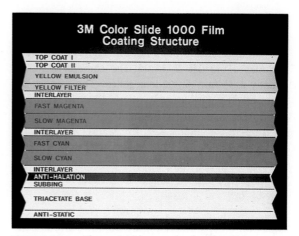

FIGURE 36-8. The typical layers of color film. *3M*

slides has given manufacturers the challenge to design and produce some useful equipment.

COPY STANDS. These units are designed to securely hold a camera directly above the image being photographed, Figure 36-9. Normally, flat

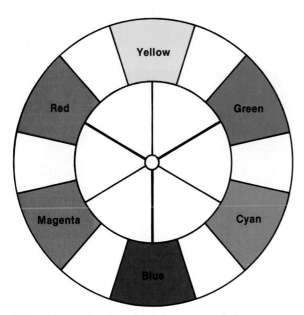

FIGURE 36-7. A color wheel shows the relationships between additive and subtractive colors.

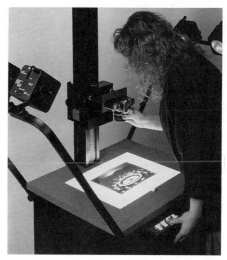

FIGURE 36-9. A copy stand is used to hold a camera in position when photographing flat, two-dimensional copy. *Bencher, Inc.*

copy from books, magazines, and artwork is photographed for use as slides. Lights on both sides provide the needed illumination. The vertical stand includes a bracket to fasten the camera, which is usually a 35-mm single-lens reflex. The camera can be moved up and down to obtain the correct image size. Auxiliary lenses frequently are attached to the camera (see unit 13). Copy stands are available in standard table model to heavy duty floor models, Figure 36-10. Many different designs, including computer controlled units, are available (see Figure 44-14).

LENSES. Macro lenses are useful when taking slides on a copy stand, Figure 36-11. Various combinations permit a wide variety of copy sizes and

types to be photographed. This type of photography is often referred to as *macrophotography.* It generally means that photographs can be made in a 1 to 1 ratio (same size) and up to 50 times the size of the original subject. See unit 13 for a review of supplemental lenses and accessories.

FIGURE 36-11. A macro lens system permits flexibility on the copy stand for macrophotography. *Olympus Corporation*

FIGURE 36-10. Two typical copy stands suitable for amateur (top) and professional (bottom) photographers. *Charles Beseler Company*

FIGURE 36-12. Color print made from a negative exposed with a pinhole camera described in unit 7. *The Pinhole Camera Company*

SUNLIGHT — NO FILTER

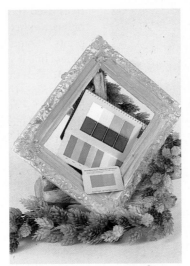

**TUNGSTEN LIGHT (3200°K) —
NO FILTER**

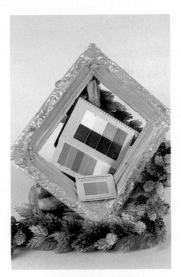

**TUNGSTEN LIGHT —
#80A FILTER**

**FLUORESCENT LIGHT —
NO FILTER**

**FLUORESCENT LIGHT —
FLUORESCENT FILTER**

FIGURE 36-13. Results of using daylight-balanced color film in tungsten and fluorescent light. The color temperature of the light does make a significant difference in capturing the true colors in the picture.

FIGURE 36-14. Standard color cards are valuable guides to creating quality color prints.

TAKING PICTURES IN COLOR

Picture taking with color film is basically the same as with black-and-white film. The major difference involves exposure control. Color film is less forgiving than black-and-white film. This makes it critical to measure the available light more accurately or to provide artificial lighting in the appropriate amounts.

1. Select the equipment, including the camera and accessories. Any camera that can expose black-and-white film can be used for color film, Figure 36-12.

2. Obtain film and load the camera. For slides, select film that has "chrome" as a suffix to its name

or is clearly marked, "for slides." For color negatives used to make prints, select film that has "color" as a suffix to its name or is clearly marked, "for prints." Some film can be used for both prints and slides, but it is not popular in the amateur market.

3. Determine the correct lighting for the selected film. Color film is balanced for either daylight or for tungsten light. Better still, obtain film that is balanced for the available light where the pictures will be taken. Light can be adjusted to match the color balance of the film as presented in unit 17. Example color photographs are shown in Figure 36-13. Electronic flash produces light similar to sunlight; thus, daylight balanced film should be used with this source of lighting.

4. Expose the film. For self-processing, it is valuable to take a picture of a standard color card, Figure 36-14. It is important to take this picture under the exact lighting conditions that the remaining pictures will be taken. Color prints can then be color balanced more quickly and accurately.

5. Process film and prints. Exposed film should be processed as soon after exposure as possible. The latent image is adversely affected by heat and humidity. Commercial processing mailers should never be placed in an outdoor, metal mailbox where high temperatures may occur during hot weather.

6. Enjoy the results. Color slides and prints provide considerable enjoyment for photographers and general viewers alike.

SUMMARY

Taking pictures in color is almost the same as taking them in black-and-white. The major difference is the concern for lighting. It is important to use color film that is balanced for the available light. Both positive-acting (for slides) and negative-acting (for prints) films are based upon the theory of subtractive colors. Finally, color films contain three light-sensitive emulsions that work in unison to capture and reproduce the vivid colors of nature and man-made products.

REVIEW QUESTIONS

Answer these questions to test your knowledge of the unit content.

1. Where is color?
2. T/F Color film first became available to the general public during the 1920s.
3. T/F Sunlight intensity has a bearing on how color film records color accurately.
4. Name the three primary light colors.
5. Which two subtractive colors when combined will produce green?
6. Which color is the complement of green?

   A. Red   C. Magenta
   B. Cyan   D. Blue

7. T/F Positive-acting and negative-acting films contain the same three basic emulsion layers.
8. Color films both before and after exposure should be stored in facilities having humidity no higher than _____?

   A. 50%   C. 25%
   B. 10%   D. 70%

9. Copy stands are most often used for a category of photography known as _____ .
10. What does color balance mean?

UNIT 37

PROCESSING COLOR SLIDE FILM

OBJECTIVES   *Upon completion of this unit, you will be able to:*
- *Prepare for color slide film processing.*
- *Process color slide (reversal) film.*
- *Explain push and machine processing of color slide film.*
- *Mount and analyze color slides.*

KEY TERMS   *The following new terms will be defined in this unit:*

Color Developer	Reversal Bath	Slide Mounter
E-6 Process	Reversal Film	Stabilizer
First Developer	Slide Film	

INTRODUCTION

Processing color slide film is a satisfying darkroom activity. It leads to colorful results quickly and economically. The roll of film that is exposed one hour can be a finished photographic product the next. Slide film, often called *reversal film* or transparency film, is well known for producing vivid colors, Figure 37-1.

Color *slide film* processing can be done either by hand or machine. Some slide films must be processed by machine due to special characteristics of the three color emulsions. Slide films that can be processed by hand are clearly identified. The most common process is matched to the Kodak Ektachrome system. It is called *Ektachrome Process E-6*, but is frequently shortened to *E-6*. All films are clearly marked on the film magazine or cartridge with the type of processing required, Figure 37-2. Several companies market E-6 chemistry kits. Four common kit names are Kodak Ektachrome

FIGURE 37-1.   Slide film (also called reversal film) is well known for producing image clarity and accurate color. (Note: This 2 × 2 (5 × 5 cm) slide was enlarged 200%.) *E. A. Dennis*

Process E-6, Photocolor Chrome-Six, Unicolor E-6, and Beseler E-6, Figure 37-3.

FIGURE 37-2.   Color slide films compatible with the "Kodak E-6 Processing" are clearly identified.

PREPARING FOR COLOR FILM PROCESSING

Being prepared for color film processing is very important. Timing and temperature are most critical in that there is limited variance in both factors. Every effort must be made to maintain the stated time and temperature for each step in the total process.

1. Obtain the chemistry. It is best to purchase one of the ready-made kits that contains all needed chemicals. Most kits have

FIGURE 37-3. Typical E-6 color slide processing kits.

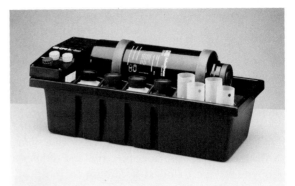

FIGURE 37-4. A complete color film processing station, including motorized film tank agitation and chemistry temperature control. *Jobo Fototechnic, Inc.*

enough chemistry to process several rolls of film.

2. Assemble the needed equipment. The standard black-and-white film processing tools and equipment will serve very well. These include:

    1 — 16-oz one- or two-roll tank
    8 — 16-oz or 500-ml plastic graduates
    8 — 16-oz or 500-ml brown, plastic
        storage bottles
    1 — stirring rod
    1 — thermometer (must be very accurate)
    1 — 10″ × 12″(25.4 × 30.5 cm) print
        processing tray
    1 — timer (must be very accurate)

Complete processing stations are available that include the film tank, storage bottles, and graduates, Figure 37-4. Two major advantages of these units are motorized agitation of the film tank and chemistry temperature control. Some units include automatic timing control.

3. Load film in processing tank. This is done exactly like black-and-white film. Two critical points to remember: (1) load the film in total darkness, and (2) do not touch the image area of the film while loading it in the tank.

4. Mix and label chemistry. Follow the instructions included with the processing kit.

Each brand contains specific chemicals and requirements. Generally, the chemicals include a first developer, reversal bath, color developer, conditioner, bleach, fixer, and stabilizer.

5. Establish chemistry temperature. After mixing, each of the chemicals must be brought up to the required temperature. Generally, this is 32 °F (17.8 °C) higher than that used for standard black-and-white film. The temperature most frequently required is 100 °F (37.8 °C) with a tolerance of ± 0.5 degree for the two developers. Temperatures of the chemicals for the other steps can generally vary between 90 and 109 °F (32.2 to 42.8 °C). Set the mixed chemical bottles in a tray of hot running water or use a temperature controlled processing station. Use a thermometer to monitor the temperature.

PROCESSING COLOR SLIDE FILM USING THE E-6 PROCESS

The processing instructional steps must be followed very closely to achieve quality results. The entire series of steps should be studied before beginning with the first developer. This will help instill confidence in completing the entire series of processing steps. The total processing time except for drying ranges between $27\frac{1}{2}$ to 44 minutes. The length of time depends on the company brand of the processing kit, and minimum and maximum times given for each step.

The following steps are provided as a general procedure for processing color slide film with the E-6 process. Specific times and temperatures must be followed according to the processing kit instructions.

1. Pour in the *first developer*. Tap the tank against a counter top to remove air bubbles clinging to the film. Agitate the tank slowly and evenly on a regular basis throughout the first developer time of 7 minutes. Keep the temperature of the developer at 100 °F (37.8 °C). This developer acts like a black-and-white developer and turns the exposed silver halides to black metallic silver. This step is the most important in the total series of processing steps. Empty the developer.

2. Add stop bath and agitate continuously for 30 seconds. The weak acetic acid solution stops the action of the alkaline base developer immediately. Empty the stop bath.

3. Refill the tank with tap water. Make certain the temperature is between 92 and 102 °F (33.3 and 38.9 °C). Water temperature that is too low or high could cause reticulation in the film emulsion. Run the water into the tank for 1 to 2 minutes. Empty the water.

4. Pour in the *reversal bath.* Tap the tank and agitate in the same manner as with the first developer. The reversal bath performs a critical task by turning the unexposed silver halide crystals into a latent image. In effect, the unexposed silver halides are chemically exposed. In some reversal processes, a bright light is used to create the latent image. Empty the reversal bath at the end of 2 minutes.

5. Remove the top from the film tank. This will make it easier to add and remove the remaining chemicals. The film is no longer light-sensitive; thus, it is completely safe to work in a lighted room.

6. Add the *color developer*. Follow the same agitation procedure as with the previous developer and reversal bath. The color developer converts the exposed silver halides to a dye. Actually, there will be the three dye colors of yellow, magenta, and cyan. The dyes produce the colors forming the positive image when light is shined through the slide film. Dye (color) is formed only in those areas where no metallic silver formed during the first development. At this point, the film has a dense black appearance. This is due to the black metallic silver that covers the image dyes. Empty the color developer after a 6-minute time period.

7. Pour in the conditioner. Agitate for the first 30 seconds and then let the film soak for the remaining $1\frac{1}{2}$ minutes. The conditioner serves as a stop bath to neutralize the action of the color developer. It also prepares the silver halides for the next step. Empty the conditioner after 2 minutes.

8. Refill the tank with the bleach solution. The purpose of the bleach is to convert the metallic silver and other waste products into dissolvable compounds. Agitate the film tank every 30 seconds during the bleach period of 7 minutes. Empty the tank.

9. Add the fixer. Handle this step exactly as with black-and-white film. The film should be immersed in the fixer for 4 minutes. The fixer dissolves the bleached silver halides, leaving only the positive dye image layers. The film now has its full color. Empty the tank.

10. Wash the film in running water for a minimum of 6 minutes. This will remove all excess chemistry that may be clinging to the surface of the film. Empty the tank.

11. Pour in the *stabilizer* solution. This helps to harden the color dyes and reduces the chance of water spots on the film. Agitate initially then let the film soak for the remaining 1 minute of time.

12. Dry the film. Carefully remove the film from the processing tank and reel. The dye image layers are soft and can be scratched easily. Squeegee the film, attach film clips, and hang to dry. It is best to use a drying cabinet with filtered and heated air. In any case, make certain the film is kept in a dust-free environment until it is completely dry.

13. Clean up. Return chemistry to the bottles and cap them securely. Wash, dry, and return equipment and tools to their established locations. Wipe up solution and water spills. Color processing chemistry can be very caustic.

PUSH PROCESSING

Slide (reversal) films produce the best results when they are exposed according to their normal speed ratings. Every effort should be made to use the appropriate ISO (DIN) rated film for the given lighting conditions. When necessary, slide films that can be processed with E-6 chemistry can be pushed to another speed.

It is important to remember that pushing means to underexpose film. For example, an ISO 200 (DIN 24) rated film can be exposed at an EI (exposure index) 400. This reduces the amount of light by one-half. This makes it possible to increase the f-stop or shutter speed by one full setting. Pushing film is useful for lower light exposure conditions.

E-6 PUSH PROCESSING	
First Development Adjustments	
Increased Film Speed	*Time Increase*
Double (2×)	2 minutes
Triple (3×)	5½ minutes
Quadruple (4×)	10 minutes

FIGURE 37-5. First development adjustments when slide film is exposed at higher ISO (DIN) speeds than its regular rating.

Push processing is completed in the same manner as regular processing except for the first developer. The first development time must be increased accordingly, Figure 37-5. It is also possible to keep the first development time the same as for regular exposure and increase the developer temperature. This is often impractical due to the very high temperatures.

MACHINE PROCESSING

High-volume film processing demands the use of automatic equipment. Most film processing equipment is designed for a specific chemical process such as for E-6 chemistry. It is possible with some equipment to change chemistry, thus making the equipment more versatile.

A typical roll-film processor for E-6 chemistry is shown in Figure 37-6. It is a daylight-operated

FIGURE 37-6. An automatic color film processor for E-6 chemistry. *Hope Industries, Inc.*

machine capable of processing one 20-exposure roll of approximately 4 feet (1.2 m) up to a series of rolls spliced together totalling 400 feet (121.9 m). The schematic diagram shows the various chemical tanks or compartments that the film passes through, Figure 37-7.

Film processors are designed to replenish the chemistry. Automatic sensors determine when new chemistry should be added so the chemistry is kept to full strength. The film enters the processor from a magazine that must be loaded in total darkness. After going through all of the chemistry, the film passes into the drier section. Many times this is referred to as "dry-to-dry" processing.

MOUNTING SLIDES

Most reversal films are mounted into 2" × 2" (5 × 5 cm) slide frames. That is the obvious reason for calling the film slide film. There are many types of slide frames available. They are made of cardboard, plastic, or glass with a thin metal frame. Cardboard and plastic mounts are the most popular because of their low cost.

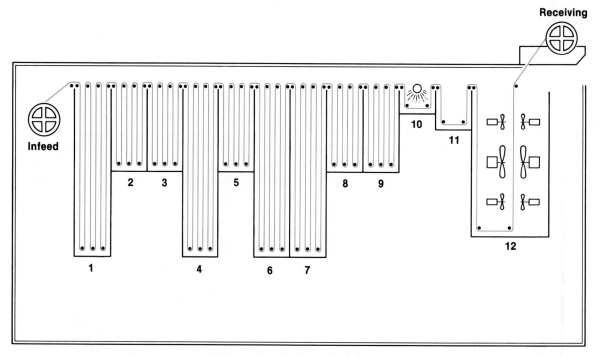

Times for Various Steps:

1. Developer 1	6 min.	7. Fixer	6 min.
2. Wash 1	3 min.	8. Wash 2	3 min.
3. Reversal	3 min.	9. Wash 3	3 min.
4. Color developer	6 min.	10. Spray wash	0 min. 45 sec.
5. Conditioner	3 min.	11. Stabilizer	1 min. 30 sec.
6. Bleach	6 min.	12. Dryer	2 min. 30 sec.

FIGURE 37-7. A schematic diagram of the film processor shown in Figure 37-6. *Hope Industries, Inc.*

Slides can be hand-mounted, semi-machine mounted, or automatic machine mounted. Some cardboard mounts contain contact adhesive. With the cut film frame in place, the frame is folded together. The two pieces stick together holding the colorful image in place. Some plastic mounts snap together while others permit the film to be slid into the frame from one side. The film emulsion is positioned to the back side of the slide frame.

Small slide-mounting machines are available that provide both heat and pressure. These are useful for cardboard frames that have a heat-sensitive adhesive. Automatic slide mounting machines are used for high volume, Figure 37-8. Units of this type can mount and label 6000 slides per hour.

ANALYZING SLIDES

Finished slides should be reviewed and analyzed to determine what can be done to improve them. A convenient method of analyzing slides is to place them in an archival plastic file, Figure 37-9. This permits a comparison of several slides at one time. The true test is to show them on a large screen with a slide projector containing a bright light.

It is important to determine the cause of an identified problem. Further, a solution must be found for each problem. Making the same mistake on a repetitive basis is a waste of time, materials, and money. The information contained in Figure 37-10 should help solve slide-making problems.

SUMMARY

Color slides provide a colorful way of capturing bright images on film. They also permit images to be viewed at same size or greatly enlarged. Processing color slide film is an extension of processing black-and-white film. The critical points to remember are to maintain the specified temperature of the chemistry and abide by the specified times for each chemical step. The results will be a colorful tribute to careful work habits.

FIGURE 37-8. An automatic mounting machine capable of mounting and labeling 6000 slides per hour. *Byers Photo Equipment Company*

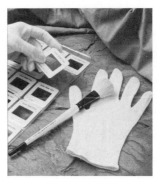

FIGURE 37-9. Keeping 2″ × 2″ slides in an archival plastic file makes it convenient for quick viewing. White, lintless gloves and a soft brush help keep the slides clean. *University Products, Inc.*

ANALYZING COLOR SLIDES		
Problem	**Cause**	**Solution**
Too dark or totally black	Film underexposed. OR Too little first development due to chemistry being too cold, insufficient time, insufficient agitation or exhausted developer.	Check camera for proper shutter operation. Also, check light meter. OR Carefully follow processing instructions.
Too light or totally clear	Film overexposed in camera or when being loaded or removed from the camera. OR Too much first development due to chemistry being too warm, too much development time, or developer too strong. Stop bath exhausted.	Check camera for proper shutter operation. Also, check light meter. Handle the film carefully. OR Check first developer temperature mixture, and specified time. Test timer for accuracy. Mix a new batch of stop bath.
Shadows too strong	Light meter not giving accurate information. Bright back lighting giving light meter false information.	Check batteries in camera. Obtain a light meter reading by getting up close to the subject. Use fill-in flash.
Light streaks	Film struck by light during loading or unloading from camera. OR Chemistry not properly agitated during first and second developers. Air pockets on film.	Load and unload film in subdued lighting. OR Follow agitation instructions very closely. Tap tank on counter immediately after adding developers to remove air pockets clinging to the film.
Fuzzy and blurred images	Incorrect lens focus. Camera moved during exposure	Focus carefully and hold the camera still while squeezing the shutter.
Greenish-bluish tint	Film subjected to excessive heat, humidity, or used after expiration date OR Daylight balanced film used indoors under fluorescent lighting	Keep film cool and dry before, during and after exposure. Observe and use film before expiration date specified by manufacturer. OR Use a Fluorescent filter (slightly magenta) to absorb the excess blue-green cast to the color of the light.
Warm yellow tint	Daylight balanced film used indoors under tungsten or photoflood lamp illumination	Use film balanced for tungsten lighting or use the correct color compensating filters.

FIGURE 37-10.   This problem-cause-solution chart is useful in analyzing the results of taking and processing color slides.

REVIEW QUESTIONS

Answer these questions to test your knowledge of the unit content.

1. The Kodak Ektachrome Process E-6 is most frequently referred to as:

A. Process E-6   C. E-6 Process

B. E-6   D. Kodak Process

2. What is the purpose of the print processing tray when hand-processing color slide film?

3. T/F A film processing tank used for black-and-white film also can be used for processing color slide film.

4. The most frequently used temperature for hand, color slide film processing is:

A. 68 °F (20 °C)   C. 100 °F (37.8 °C)
B. 85 °F (29.4 °C)   D. 118 °F (47.8 °C)

5. What does the first developer do when processing color slide film?

6. Which chemical solution used in color slide film processing does the same thing as a bright light?

7. T/F During processing, color slide film is immersed in the stabilizer solution prior to the fixing solution step.

8. Push processing of color slide film requires adjustment(s) in how many chemical solution steps?

A. 1   C. 3
B. 2   D. 4

9. T/F Machine processing of color slide film requires a minimum of four rolls of film to be processed at one time.

10. Which slide mounts are the most expensive?

A. Cardboard   C. Paper
B. Plastic   D. Glass

UNIT 38

PROCESSING COLOR NEGATIVE FILM

OBJECTIVES   *Upon completion of this unit, you will be able to:*
* *Prepare to process color negative film.*
* *Process, handle, and store color negative film.*
* *Describe machine processing of color negative film.*
* *Analyze color film negatives.*

KEY TERMS   *The following new terms will be defined in this unit:*
C-41 Process
Processing Kit

INTRODUCTION

Color negative film contains three emulsion layers—yellow, magenta, and cyan, Figure 38-1. Each layer is sensitive to one of the three primary light (additive) colors. Yellow is sensitive to blue, magenta is sensitive to green, and cyan is sensitive to red. Once exposed to light, silver is produced in each of the three layers during development. The image highlights will be dense, while the shadows will be light. Also during development,

the emulsion layers will acquire their negative color dye images.

PREPARING TO PROCESS COLOR NEGATIVE FILM

Several brands of *processing kits* are available for hand processing color negative film, Figure 38-2. The results of each brand vary slightly regarding color balance, density, and contrast, but all

KODAK VERICOLOR III PROFESSIONAL FILM 6006, TYPE S

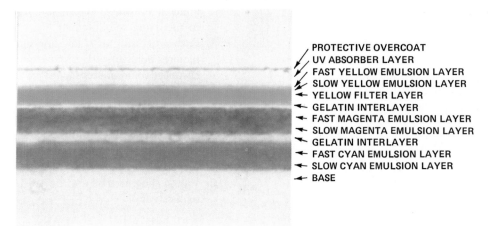

PROTECTIVE OVERCOAT
UV ABSORBER LAYER
FAST YELLOW EMULSION LAYER
SLOW YELLOW EMULSION LAYER
YELLOW FILTER LAYER
GELATIN INTERLAYER
FAST MAGENTA EMULSION LAYER
SLOW MAGENTA EMULSION LAYER
GELATIN INTERLAYER
FAST CYAN EMULSION LAYER
SLOW CYAN EMULSION LAYER
BASE

FIGURE 38-1. A cross-sectional photomicrograph of color negative film. Without magnification, the yellow emulsion layer is 1/10 the thickness of a human hair. *Eastman Kodak Company*

duce acceptable results. Some kits have six steps, while others require eight steps from predevelopment through drying. Power concentrates are used in a few kits, but most use liquid concentrates that are mixed with water. Tap water can be used, but if there is any doubt of contamination, distilled water should be used.

1. Obtain the chemistry. Select a ready-prepared kit that includes all needed chemical concentrates. Kits also contain complete instructions. The industry standard color negative process is called *C-41*. It was designed by Eastman Kodak for their own color negative films. Other companies have designed their chemistry and films to match these standards except for a few exceptions.

2. Gather the needed equipment. Standard black-and-white film processing tools and equipment will serve well.

4 — 16 oz or 500 ml plastic graduates
4 — 16 oz or 500 ml brown plastic
storage bottles

FIGURE 38-2. There is a wide choice of processing kits for color negative film.

1 — stirring rod
1 — thermometer, high-accuracy
1 — 10″ × 12″(25.4 × 30.5 cm)
print processing tray
1 — timer, high-accuracy
Temperature-controlled processing
stations are self-contained and convenient to

FIGURE 38-3. A complete processing station for small volume C-41 color negative processing. *Jobo Fototechnic, Inc.*

use, Figure 38-3. These units take the place of all of the above listed items except for a timer. Some units, though, do contain a timing mechanism that controls agitation and length for each chemical solution step.

Preparation steps 3, 4, and 5 are the same as for processing color slide film (see unit 37).

PROCESSING COLOR NEGATIVE FILM USING THE C-41 PROCESSING KIT

Color negative film kits contain specific instructions on how the chemistry solutions should be used. Every effort must be made to closely follow the step-by-step instructions. Chemists have formulated each solution to perform a specific part of the total process. Making chemical substitutions is not advisable.

The following procedure will serve as a general guideline for using any one of the several C-41 processing kits for color negative films. Use it accordingly.

1. Check the temperature of the developer. It must be within the parameters as specified in the kit instructions. Processing temperatures can vary from 65 °F (18.3 °C) to 105 °F (40.5 °C).

2. Add the developer to the tank as rapidly as possible. Agitate slowly and continuously for the first 15 to 30 seconds. Complete the developing time by agitating for 2 seconds every 15 seconds. Developing times vary from $2\frac{3}{4}$ to 18 minutes depending on the temperature and kit brand. Remember to begin draining the developer solution from the tank 10 to 15 seconds prior to the total developing time. Extending the developing time while draining the tank can cause overdevelopment of the negatives.

3. Add the bleach bath immediately. Agitate continuously for 15 to 30 seconds and then for 2 seconds every 30 seconds for the remainder of the time. Bleach stops the developer action immediately and causes a chemical reaction. It changes the metallic and colorless condition. If this was not done, light could not pass through the color emulsion layers of the film. Empty the bleach after a $6\frac{1}{2}$ minute time period.

4. Remove the top from the film tank. The film is no longer sensitive to light. This will make it easier to add and remove the remaining chemicals.

5. Wash the film for 3 to 4 minutes with tap water. Make certain the temperature is within tolerance levels so reticulation is avoided. The overflow washing method helps to remove previous chemicals quickly when the wash water rises from the tank bottom, see Figure 27-6.

6. Pour in the fixer. Use the same agitation technique as with black-and-white film. The unused silver halides are dissolved by the fixer. Empty the fixer after $6\frac{1}{2}$ minutes.

7. Wash the film as in step 5. This time the water removes the remaining fixer from the film surface.

8. Add the stabilizer solution. It serves to harden the three-layer dyes and helps to reduce water spots on the film. Keep the film in the stabilizer for about $1\frac{1}{2}$ minutes.

9. Dry the film. Handle it in exactly the same way as with black-and-white film. The film will have a multiple-color negative appearance.

10. Clean up the work area. Save chemicals for processing other rolls. Remember that color developers have a short life after being mixed to working solutions.

HANDLING AND STORING NEGATIVES

Color negatives and black-and-white negatives are handled and stored in the same way. Always handle them by the edges. White, lint-free gloves help protect negatives from skin abrasions, moisture, and dirt from the fingers.

1. Cut the negatives into 5 or 6 exposure frame lengths. Use a sharp clean scissors and cut in the center between two exposures.

2. Place each color negative strip in a protective sleeve. Both single or multiple-page sleeves work well.

3. Label the individual sleeves or envelope in which the sleeves are placed. This will help in locating the negatives at a later date.

4. File the negative envelope in a clean, dry location.

MACHINE PROCESSING

Film processing laboratories must use high-speed equipment. Also, the equipment must be very reliable so that every roll of color negative film is processed exactly the same way. A typical C-41 film processor is shown in Figure 38-5. It is capable

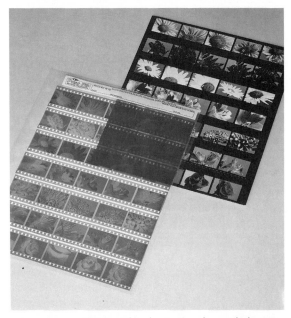

FIGURE 38-4. Using archival negative sleeves helps protect color negative strips from gathering dust and finger prints. Contact proof sheets assist the photographer when selecting negative frames for printing. *Print File, Inc.*

FIGURE 38-5. A C-41 high-speed color negative film processor. *Hope Industries, Inc.*

FIGURE 38-6. A manual color film, C-41 processor that is designed to maintain constant chemical temperature along with pushbutton control for each processing step. *Richcolor Systems, Inc.*

of processing 35-mm color negative film with five strands (film widths) across with an effective output of 9.2 ft (2.8 m) per minute.

Smaller color laboratories often use tank processing systems, Figure 38-6. These systems can be used to process a few rolls to hundreds of rolls of color negative film per week. Units of this type give consistent temperature control and make it convenient to use each of the several chemical solutions.

ANALYZING COLOR NEGATIVES

After processing, color negatives should be analyzed. This provides an opportunity to identify problems, causes, and solutions, Figure 38-7. Improvements can best be made when a systematic procedure is followed.

1. Identify the problem.
2. Isolate the cause(s).
3. Propose solutions.

ANALYZING COLOR NEGATIVES		
Problem	**Cause**	**Solution**
Too dark or totally black	Film overexposed OR Developed too long. Bleach and/or fix exhausted	Check camera shutter operation. Also, check light meter. OR Re-read processing instructions. Check age of chemistry.
Too light or totally clear	Film underexposed OR Developer weak	Check camera shutter operation. Also, check light meter. OR Use new developer solution.
Dark streaks	Film struck by light during loading or unloading from camera	Load and unload film in subdued lighting.
Clear streaks	Film touching together in tank during processing. Developer could not circulate freely	Carefully load film into the processing tank.
Spots	Chemistry not properly agitated during processing	Follow agitation procedures very closely during each processing step.

FIGURE 38-7 This problem–cause–solution chart is useful in analyzing the results of taking and processing color negative film.

4. Determine the best solution.
5. Complete the process again.

SUMMARY

Processing color negative film is an exciting part of continuous-tone photography. The procedure is very similar to processing black-and-white film. The one major difference is the required higher temperature of the chemical solutions. Some color negative processing kits do permit lower temperatures, but the total processing time is usually much longer. After processing, the negatives are used to make full color prints.

REVIEW QUESTIONS

Answer these questions to test your knowledge of the unit content.

1. Name the three color sensitive layers of color negative film.
2. T/F Liquid chemistry is used in all color negative film processing kits.
3. How many chemical solutions are used in the C-41 process?

   A. 1   C. 3
   B. 2   D. 4

4. T/F Chemical substitutions can be made in color processing kits without fear of causing problems.
5. Which step in color negative film processing is significantly different from black-and-white film processing?
6. Which processing chemical solution is designed to reduce water spots on color film?

   A. Bleach   C. Stabilizer
   B. Fixer   D. Stop bath

7. T/F Color chemistry has a short storage life once mixed into a working solution.
8. How are color negatives best stored?
9. T/F Color processing laboratories use high-speed film processing equipment for all of their work.
10. What is the first step in analyzing color negatives?

CREATING COLOR PRINTS

OBJECTIVES   *Upon completion of this unit, you will be able to:*
- *Describe color photographic paper.*
- *List equipment and darkroom needs for color printing.*
- *Prepare for exposing and processing color prints.*
- *Expose and process color prints.*

KEY TERMS   *The following new terms will be defined in this unit:*
Analyzer
Dichroic Enlarger
Filtration

INTRODUCTION

Making prints is an exciting part of photography. This is especially true in color photography, Figure 39-1. Color prints can easily be made from both color negatives and color slides. The procedure for each is basically the same. The major differences include the *filtration* used in the enlarger for color balance, the type of photographic paper, and print processing.

COLOR PHOTOGRAPHIC PAPER

Photographic paper for making color prints contains the same basic layers as does color film. The main difference is the substance. Another difference is the order of the color emulsion layers in standard print paper, Figure 39-2. The blue-sensitive layer (yellow color emulsion) is placed next to the paper base. This provides a more accurate color balance when light is reflected from the print.

Many photographic papers are available for making color prints. There are two basic categories of color print paper. One is for making prints from negatives, and the other is for making prints from slides. Both categories of paper are manufactured with utmost precision to give accurate color reproduction. It is necessary to use the correct paper for the type of film image being used.

FIGURE 39-1. A quality color photograph of a pleasant scene is always nice to view. *E. A. Dennis*

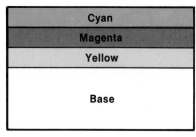

Yellow
Magenta
Cyan
Base

Color film

Cyan
Magenta
Yellow
Base

Color paper

Yellow emulsion = Blue sensitive
Magenta emulsion = Green sensitive
Cyan emulsion = Red sensitive

FIGURE 39-2. The order of the color emulsions are different in color film and standard color paper.

EQUIPMENT AND DARKROOM NEEDS

Standard black-and-white processing equipment and tools can be used for several steps in making color prints. Some special equipment is necessary, including a color enlarger, filters, analyzer, temperature control, and processing unit.

ENLARGER. An enlarger with a good quality lens is very important. The sharpness of the finished print whether it be black-and-white or color is almost totally due to the lens. Also, color prints can only be made with an enlarger that can project the subtractive colors of light. This can be done in one of two ways. First, a color enlarger includes a built-in filtering system, Figure 39-3. It permits convenient settings for different values of yellow, magenta, and cyan filtration. These are often called *dichroic enlargers* when glass filters and a mirror are used to project the selected color of light. Economical color enlargers are designed to use plastic filters. The best color enlarger head contains three separate light sources instead of the usual one.

Microprocessors (computers) are used with some color enlargers, Figure 39-4. These units help to determine precise exposures based upon the colors in the color negative or slide. Digital readout

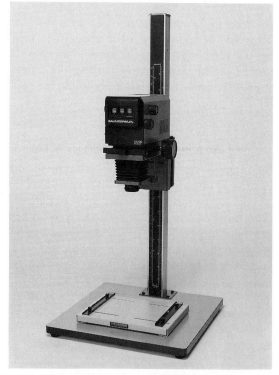

FIGURE 39-3. A condenser-type, medium format dichroic enlarger that can be used to expose quality color prints. *The Saunders Group*

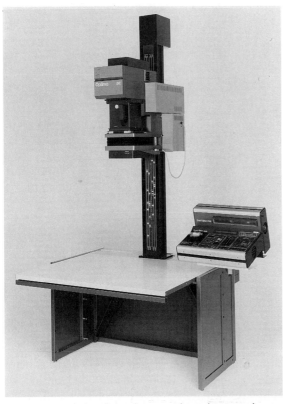

FIGURE 39-4. A color enlarger with sophisticated computer controls helps improve the accuracy of the initial color print exposures plus the repeatability of exposures is very accurate. *Colenta America Corporation*

FIGURE 39-5. A color printing filter set used to convert a black-and-white enlarger to a color enlarger. *Charles Beseler Company*

makes it convenient to see the exact filtering values selected by the computer.

FILTERS. Black-and-white enlargers can be converted for making color prints. This is done by adding filters, Figure 39-5. Filter sets are made of fade-resistant, tough plastic in various sizes from 3″ × 3″ (7.6 × 7.6 cm) to 6″ × 6″ (15.2 × 15.2 cm). The yellow, magenta, and cyan filters vary in density, making it possible to select the correct combination for a given negative or slide. They are numbered according to their density. A typical color printing set consists of 20 to 25 filters. They range in density from 2.5 (low density) to 50 or more (high density).

The least number of filters possible should be used at one time. For example, two 10Y (yellow) filters are the same as one 20Y, but it is always better to use one instead of two. Only two different filter colors should be used at one time. Using all three subtractive colors at once will produce a neutral gray to black.

Filters are added or subtracted depending on whether color prints are being made from negatives or slides. A specific color cast in a print from a negative requires added filtration of the color. A print from a slide requires less filtration of the color. The amount of filtration to add or subtract can be determined by looking at prepared test prints.

Filters should be located between the enlarger lamp and the film holder for best results, Figure 39-6. Many enlargers have a filter drawer or opening that accepts filters. It is possible to position filters below the lens, but this is not the ideal location. Sometimes the sharpness of the projected image is affected.

One filter must always be placed in a black-and-white enlarger converted for color. That is an

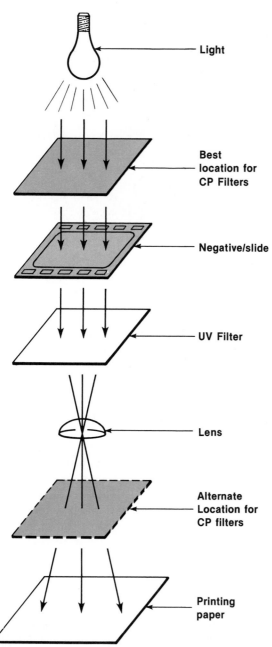

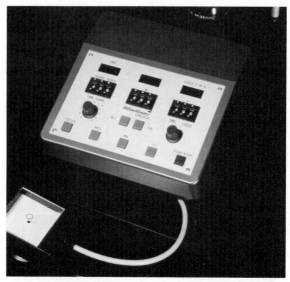

FIGURE 39-7. A color analyzer is an extension for the human eye. *Labex Engineering Corporation*

FIGURE 39-6. The ideal and alternate locations for color printing filters in an enlarger.

ultraviolet (UV) filter. It removes UV light coming from the enlarger lamp. Uncontrolled ultraviolet light rays will cause severe problems with color balance.

ANALYZER. A color *analyzer* is an electronic tool that serves as an extension of the human eye, Figure 39-7. The main purposes of an analyzer are to speed the making of color prints, establish image color accurately, and save on print paper material waste. The enlarger shown in Figure 39-4 includes a color analyzer as part of its total systems. A color analyzer is very useful but is not necessary for making amateur color prints.

TEMPERATURE CONTROL. The same type of temperature control units are needed for color print processing as for color film processing. These can be very elaborate or as simple as a tray with warm running water.

PROCESSING UNIT. Paper prints can be processed in trays, but rotary processors containing drums work much better, Figure 39-8. These are similar to film tanks except they are larger and will

FIGURE 39-8. A rotary type print processor that permits accurate temperature control, accurate drum rotation speed control, and convenient chemical handling. *Jobo Fototechnic, Inc.*

accept paper sizes from 4″ × 5″ to 16″ × 20″ (10.2 × 12.7 cm to 40.6 × 50.8 cm). Processing drums allow good temperature control and agitation of the chemical solutions. Other hand- and power-operated processors are suitable for amateur color printmaking.

Color printmaking is best accomplished in a separate darkroom. It must be made totally dark. This makes it possible to keep black-and-white chemistry and materials separate from those used for color.

PREPARING TO EXPOSE AND PROCESS COLOR PRINTS

Proper preparation for exposing and processing color prints will save time and increase quality. The following six steps should be completed with care.

1. Make safety preparations. Have a pair of splash-proof goggles and a pair of thin rubber gloves available for chemical mixing and processing. Check the darkroom floor for articles that might cause someone to trip or slip and fall.

2. Prepare the enlarger. Clean and vacuum it if needed. Check all electrical connections and other mechanical functions of the unit. Ready the timer for the test exposures.

3. Establish a color reference. Select the negative or slide that was exposed of a standard color card (see Figure 36-14). If this is not available, a commercial color reference slide or negative can be used.

4. Compose the image. Position the negative or slide in the film holder, and insert it into the enlarger. Place a piece of white paper in the easel for ease in focusing the image. Focus, size, and crop the image in the same way as with black-and-white print making. The film emulsion must be down when using conventional color print paper and processing.

5. Obtain and mix the chemistry. There are several brands available for processing prints from negatives and prints from slides. It is, though, important to use chemistry compatible with the negatives or slides and with the photographic paper, Figure 39-9. Establish the proper chemical solution chemistry and follow the print processing instructions closely. Be careful not to spill the chemistry as it is highly caustic. This can cause permanent stains in clothing and counter tops.

6. Establish safelight conditions. Most color printing papers are very sensitive to all light.

MAKING TEST PRINTS

Several test printing methods are available. The conventional method includes making a test print with four different exposures. This is usually done without filters, except for the UV filter. It is helpful to use a special four-part easel that permits an exposure to one quarter of the print paper at one time. After processing, the no filter

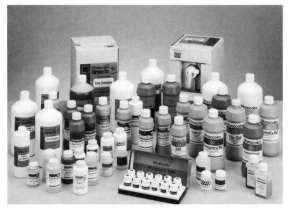

FIGURE 39-9.   Many chemicals are required when processing color film and prints. Correct chemistry selection is an important key to a successful result. *Jobo Fototechnic, Inc.*

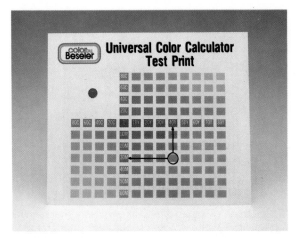

FIGURE 39-10.   A print made with a color calculator helps to determine the color printing filters and exposure. *Charles Beseler Company*

test print is evaluated to determine the best exposure length. Two more test prints with color printing filters often are needed before the correct color balance is obtained. When the correct filters and exposure times are known, prints can be exposed and processed with ease.

Test print calculators are convenient to use. They save time, material, and cost.

1. Set the enlarger lens at f/11 and the timer for 10 seconds or units.

2. Place a special light diffuser in the filter accessory holder below the lens. This distributes the image light that falls on the image area of the easel.

3. Turn out the darkroom lights. This makes it necessary to know where everything is located in total darkness.

4. Place a sheet of color enlarging paper in the easel.

5. Using a special color calculator test sheet, position it on the enlarging paper. Be sure it lies flat against the paper.

6. Make the exposure.

7. Process the test print. Use the process appropriate for the print paper being used.

8. Evaluate the test print. Locate the neutral gray square on the test print, (Figure 39-10). The two filters needed to create the correct color balance will be found horizontally and vertically from the gray square. The example in Figure 39-10 indicates that a 30M (magenta) and a 40Y (yellow) filter are needed. Also, the test print calculator provides information on the correct exposure to use when the filters are placed into the enlarger. Filters absorb some light, thus more exposure will be needed.

9. Expose and process a print based upon the test print information. Evaluate the print against the original scene. Change filtration as needed to obtain acceptable colors in the finished print.

PROCESSING COLOR PRINTS

Conventional color print processing can be as short as one step or as long as eight to ten steps.

Much depends on the brand of chemicals being used. Also, it depends on whether prints from negatives or slides are being made. The basic processing steps for color prints are listed.

Prints from Negatives	*Prints from Slides*
1. Preheat paper and tank	**1.** Preheat paper and tank
2. Developer	**2.** First developer
3. Stop bath	**3.** Wash
4. Bleach-fix	**4.** Color developer (reverse image)
5. Wash	**5.** Wash
6. Dry	**6.** Bleach-fix
	7. Wash
	8. Dry

ROTARY PROCESSING PROCEDURE

The most important aspect of color print processing is consistency. Every phase must be completed correctly according to the specific procedures made available with the chemistry and mechanical processor. Temperature control is very critical when processing color film and paper prints, thus every effort must be made to maintain chemical temperatures within an accuracy range of + or −0.36 °F (+ or −0.2 °C). The following general procedures are utilized to process color prints using a drum-type mechanical processor as shown in Figure 39-8.

1. Fill the tempering bath area where the drum rotates with water. The water level should come to the shoulders of the chemical bottles that are located in individual compartments at the front of the tray.

2. Turn on the circulating pump. This circulates the water around the chemical bottles, graduates, and processing tank.

3. Set the temperature control to the desired processing temperature. For most color chemistry, the required temperature is usually 100 °F (38 °C).

4. Mix the working chemistry from the concentrate solutions and pour into the chemical bottles located in the processing unit. The usual chemicals include the developer, stop bath, bleach fix, and rinse water. If possible, have two or three of the bottles filled with water for the pre-wetting and washing steps of print processing. The amount of each solution, including the water, depends on the processing drum size and the number of prints that will be processed.

5. Wait a period of time until the processing chemistry is up to the correct temperature. A mechanical processing unit must accurately maintain the chemical and water temperature for quality results.

6. Select the correct print drum, Figure 39-11. The drum size depends on the color print paper size.

7. Adjust the processor to accept the selected drum.

8. Load the print drum with an exposed sheet of color print paper. This must be done in a 100% darkened darkroom. Replace and

FIGURE 39-11. Print processing drums are available for a wide range of photographic paper sizes. *Jobo Fototechnic, Inc.*

FIGURE 39-14. A color photographic print should accurately represent its subject. *Eastman Kodak Company*

FIGURE 39-12. Pouring a chemical solution into the processing drum through the special funnel system of the processing unit. *Jobo Fototechnic, Inc.*

FIGURE 39-15. A beautiful scene, creative photography, and quality processing can be combined to make a lasting record. *James A. Riggs*

FIGURE 39-13. Draining a chemical processing solution from the print drum. Some solutions can be saved for reuse while others must be properly discarded after each use. *Jobo Fototechnic, Inc.*

secure the cap of the print drum before returning to the lighted processing area.

9. Install the loaded processing drum into the processing unit. Be certain the drum is correctly fastened to the drive head. It must rotate at a consistent speed plus, when lifted for draining chemistry, it must not become disengaged.

10. Fill the graduates, located in the processing unit, with the correct amount of each solution. The number of ounces or millimeters depends on the drum size. A chart included with the processing drums should be consulted to determine the amount for each solution including the water.

11. Turn on the motor to begin rotating the drum.

12. Pre-wet the print paper with tempered water. Pour the water as rapidly as possible

into the processing drum, Figure 39-12. Allow the drum to rotate for one minute and drain the water from the drum.

13. Pour in the developer solution while the drum is rotating. Depending on the developer and the process, the developing time will usually range from 2.0 to 2.5 minutes.

14. Drain the drum as rapidly as possible at the end of the developing time, Figure 39-13. The developer can be saved for future use or properly discarded.

15. Add the stop bath to the drum and rotate for 30 seconds. Drain the stop bath from the drum.

16. Pour in the bleach fix and rotate the drum for 1.5 minutes. Drain the bleach fix from the drum.

17. Wash the print with four to six changes of water at 30 second segments. Water changes are important to the removal of processing chemistry on and in the color print.

18. Remove the print from the processing drum. The wet print must be handled carefully as the image emulsion is very soft. It is best to handle it gently by the edges.

19. Remove the surface water, as with a black and white print, and place it in a drying cabinet. If the exposure and processing were correct, a useful print should be the result, Figure 39-14. Be sure the print is completely dry before removing it from the dust-free drying cabinet.

20. Quality color prints of pleasant scenes are always a joy to look at and study, Figure 39-15. The time, cost, and effort required to produce color photography is rewarding for both the amateur and professional photographer alike.

ANALYZING COLOR PRINTS

The information found in Figure 39-16 should help solve color printmaking problems. Identifying problems and determining causes lead to successful solutions.

COMMERCIAL PRINTING AND PROCESSING

Speed and quality are two important words to commercial color printmaking. Having one without the other will not prove successful. Exposure units containing microcomputers, multiple lenses, and bulk rolls of print paper help maintain speed and quality, Figure 39-17. Exposure units of this type do the same job as a quality enlarger but much faster.

High-speed printing units require sophisticated print processors, Figure 39-18. Computerized program control monitors chemistry replenishment, roller transport speed, and processing temperature. Commercial equipment of this type gives consistent results through proper operation and management.

SUMMARY

Making color prints from negatives and slides is an extension of black-and-white printmaking. Some additional equipment is needed for both exposing and processing prints. Making a color exposure requires exact timing and correct color filtration. Print processing demands close attention to using correct chemistry, temperature, and timing. Various test exposure methods and processing equipment make color printmaking a practical part of continuous-tone photography.

REVIEW QUESTIONS

Answer these questions to test your knowledge of the unit content.

1. How many color emulsion layers does photographic print paper have?
2. T/F Emulsion layers of conventional color print paper and color film are in the same order.
3. Which of the following filter colors is not included in a color enlarger?

    A. Green   C. Cyan
    B. Yellow   D. Magenta

4. T/F Microcomputers have found a use with color enlargers.
5. Why shouldn't all three colors of filters be used in an enlarger at one time?
6. Which filter should always be included in a color printing enlarger?
7. Which safelight is most frequently used when making color prints?

    A. Red   C. Yellow
    B. None   D. Blue

8. T/F Film negatives and slides are posi-

Problem:	Solution:
Print too dark.	Use shorter exposure time.
Print is too light.	Use longer exposure time.
Black areas of print are blue.	Increase developer time and/or temperature.
	First developer is old or exhausted. Mix fresh chemistry.
Dark blotches in print.	Paper has been exposed to light.
Print has light colored stripes.	Processor is not level and/or drum is floating in upper trough. Check processor with level. If drum is floating lower water level in upper trough.
	Insufficient chemistry amount. Use either amount recommended by JOBO or by chemical manufacturer, whichever is highest.
Print is light or off color on end away from processor.	Chemistry volume, insufficient or drum not level. Check if processor is level. Make sure drum is not floating.
Print has stripes from end nearest processor to end farthest from processor.	Use 1 minute prewash. Place Drum on processor more rapidly after turning horizontal.
Light spots in print.	Bleach/Fix contamination. Clean drum and cap assembly carefully.
Color shift when switching from test drum (#2820) to larger drums.	Increase amount of fresh chemistry used in processing.
White in the print is impure.	Wrong safelight in use. Darkroom is not light tight.

FIGURE 39-16. This problem-solution chart is useful in solving color print processing results. *Jobo Fototechnic, Inc.*

tioned in the enlarger emulsion up or down depending on the type of print paper.

9. The first test print is most often made —with or without—color printing filters.

10. T/F In conventional processing, prints made from slides must go through more steps than from negatives.

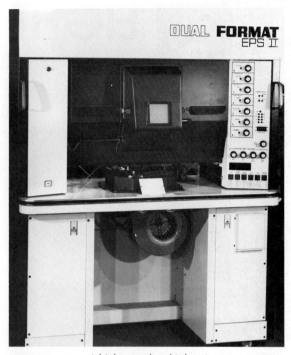

FIGURE 39-17. A high-speed multiple print exposure system. *Photo Control Corp.*

FIGURE 39-18. A color print processor that has computerized control. *Kreonite, Inc.*

CHAPTER EIGHT

PHOTOGRAPH UTILIZATION

MOUNTING PHOTOGRAPHS

OBJECTIVES   *Upon completion of this unit, you will be able to:*
* *Explain the value of mounting photographs.*
* *List the tools, equipment, and supplies used for mounting photographs.*
* *Identify the four basic mounting methods.*
* *Mount a photograph on a mounting board with dry mounting tissue.*

KEY TERMS   *The following new terms will be defined in this unit:*

Double Matting	Flush Mounting	Mounting Board
Dry Mounting Press	Mat Cutter	Pressure-sensitive Mounting Sheets
Dry Mounting Tissue	Matting	Tacking Iron
Face Mounting	Mounting	Window Matting

INTRODUCTION

Mounting helps focus attention on the content of photographic prints. Photographs can easily be displayed when securely fastened to thick cardboard-like material called *mounting board*, Figure 40-1. Creativity and skill are required to correctly mount photographs. This talent helps round out the "complete photographer."

MOUNTING TOOLS

Several hand tools commonly found in photographic laboratories and homes are used to mount prints. It is important to keep each tool in top condition. Knives must be sharp and measuring tools clean to achieve good results.

ART KNIFE.   This is used to trim prints and mounting tissue. Replaceable blades make it possible to keep a sharp art knife available at all times.

BOX KNIFE.   This type of knife is sometimes called a utility knife. This is a heavy-duty knife used to cut the thick mounting board material. Replaceable blades are easy to install and securely hold in place.

SCISSORS.   A good quality scissors is essential when cutting and trimming materials used for mounting. Wiping the blades with a paint-solvent dampened cloth will keep them free of adhesives and dirt.

FIGURE 40-1.   Mounted photographs are pleasing to view and enjoy. *Doran Enterprises, Inc.*

295

FIGURE 40-2. A T-square with large numbers and markings helps to position photographs on mounting materials. *Falcon Safety Products Inc.*

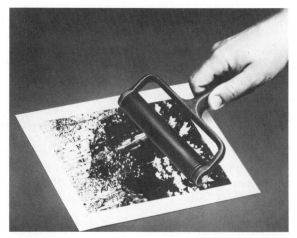

FIGURE 40-3. A roller is an excellent tool for burnishing adhesive-backed prints onto mounting material. *Falcon Safety Products Inc.*

T-SQUARE. A combination wood and plastic or metal T-square is useful in positioning prints. Some T-squares specially made for photograph mounting contain large numbers and markings for convenience and accuracy, Figure 40-2.

ROLLER. A suitable type contains a hard rubber roller secured to a wood or metal handle, Figure 40-3. It is used to burnish prints onto mounting material when spray or pressure-sensitive adhesive is used. A standard rolling pin works well for this purpose, too.

MOUNTING EQUIPMENT

Photographs can be mounted without special equipment; however, convenience and quality are improved when one or more of the listed items are used.

HEAVY-DUTY BAR CUTTER. Cutters of this type are used to cut the thick—hard mounting boards, Figure 40-4. The blade is heavy and made of hard steel. A clamp securely holds the material while it is being cut.

TRIMMER. A trimmer that uses a single-edged razor blade or a rotary knife can be used to trim thin materials, Figure 40-5. A unit of this

PENCIL. A soft-lead pencil is needed to mark distances and sizes of prints and mounting materials. A blue lead pencil permits visible markings and is easy to remove from mounting materials. Light, thin lines should always be used; thus, pencils should be kept sharpened to a point.

ERASER. A ruby-red or soap eraser is essential to remove pencil marks from the mounting materials. Smudges and dirt marks often can be removed with either of these soft erasers.

BRUSH. A drafter's brush is useful to remove erasure rubbings from mounting materials. The soft bristles will not harm the surface of photographic prints and mounting materials.

MEASURING RULE. A wood, plastic, or steel measuring rule is essential to proper sizing and positioning of mounting materials and photographs. It is important to use a rule that contains easy-to-read, accurate distance markings.

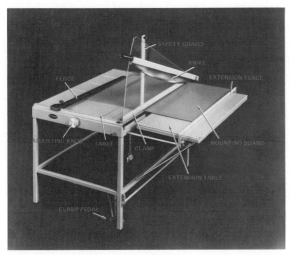

FIGURE 40-4. A heavy-duty bar cutter used to cut thick mounting material. *Michael Business Machines, Corp.*

type is especially useful when removing borders from photographs and trimming dry mounting tissue along print edges.

MAT CUTTER. Mats are pieces of mounting board with rectangular or circular openings. They are placed over photographs to improve their appearance. *Mat cutters* are available to make straight cuts or circular cuts, Figure 40-6. These special

FIGURE 40-6. A circular-type mat cutter used to cut ovals and circles in mounting boards.

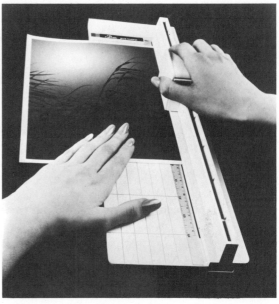

FIGURE 40-5. Using a razor-sharp trimmer to remove print borders. *Falcon Safety Products Inc.*

cutters make it possible to cut smooth, exacting openings.

DRY MOUNTING PRESS. This type of press is essential for professional mounting results, Figure 40-7. Mounting presses are used to provide even heat and pressure to the photograph, adhesive sheet, and mounting board. Quality units contain pressure adjustments, temperature control, and timers. Manufacturers make several sizes of dry mounting presses.

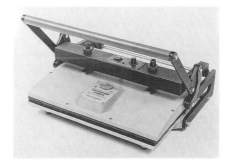

FIGURE 40-7. A dry mounting press provides an even distribution of heat and pressure. *Seal Products, Inc.*

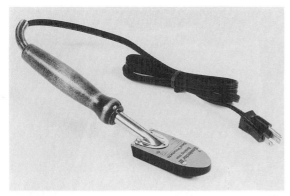

FIGURE 40-8. A tacking iron is used to adhere dry mounting tissue to photographs and mounting board. *Seal Products, Inc.*

TACKING IRON. This device contains a heating element and a temperature control within a small metal platen, Figure 40-8. It is used to adhere (tack) dry mounting tissue in small areas to photographs and mounting board. A wooden or plastic insulated handle makes it easy to use a tacking iron without getting burned.

HAND IRON. A standard pressing iron works well for dry mounting. The "cotton" setting is about the right heat for most dry mounting adhesive materials.

MOUNTING SUPPLIES

Three major supply items are needed to mount photographs on rigid material. These items are adhesive, mounting board, and protective paper.

ADHESIVE. Three basic types of adhesive are available for fastening materials together. *Dry mounting tissue* is one of the oldest and best adhering materials. It is placed between the items to be mounted. It contains a base sheet with adhesive coated on both sides. The adhesive becomes tacky when heated and adheres the materials together while being pressed together. It helps to make a professional-looking mounting. *Pressure-sensitive mounting sheets* are sometimes referred to as "cold-mounting." The adhesive is coated on both sides of a base sheet and protected with cover sheets. Photographs are adhered to mounting board with this material by removing the protective cover sheets and applying moderate pressure. *Wet adhesive* is applied to the back of a photograph. It can be applied with a brush or from a spray can. The print is then positioned and pressed (burnished) to make it adhere to the mounting board.

MOUNTING BOARD. Many different colors, surface textures, thicknesses, and sizes of mounting board are available. Mounting board is a paper-based product that contains several compressed layers. This makes the material very hard and sturdy. The best type of mounting board is free of acid content and is referred to as "archival" quality. Acids in mounting boards can cause photographic prints to discolor, become brittle, and actually fall apart. Some mounting boards contain a styrofoam center core making them about 1/4" (6.4 mm) thick but very light in weight.

PROTECTIVE PAPER. Photographs must be protected from damage while being mounted. Clean bond typing paper (white) works well for most situations. The best material to use is called "release paper." It is a special silicone-treated paper that resists adhesives, thus protecting photographs, mounting boards, and the platens of dry mounting presses.

MOUNTING METHODS

Several methods of mounting can be used to enhance the beauty of a photograph. Each method has advantages and disadvantages but does serve a specific purpose.

FACE MOUNTING. Adhering a print to a mounting board and leaving borders on all four sides.

FLUSH MOUNTING. Fastening a print to a mounting board of the same size or trimming the mounting board to match the print size after adhering.

WINDOW MATTING. Positioning a frame made from mounting board having a rectangular or circular cut opening. Generally the photograph is face mounted before the *matting* is placed on top. The opening should be smaller than the photograph so that the print borders are not visible. The window mat can be hinged to the base mounting board with special archival tape. It also can be securely adhered using one of the adhesive methods.

DOUBLE MATTING. This is an extension of the window mat method. A second mat is cut with a larger opening than the first mat. The opening size increase varies depending on the print size, but 1/4″ to 5/16″ (6.4 to 7.9 mm) on each side is common.

BASIC MOUNTING PROCEDURE

Mounting photographs can be done by several methods. First, one of the three general categories of adhesives can be used, and second, one of the four mounting methods can be used. The following steps cover *face mounting* using *dry mounting tissue.* These procedures serve as a foundation for using the other materials and methods.

SAFETY REMINDER

Cutting tools and equipment must be used with care. Remember to keep hands and fingers behind a knife while cutting and trimming. Also, equipment using heat and pressure must be treated with respect. Expect heat-related equipment to be hot until found otherwise.

1. Determine the mounting board size. A good size for 5″ × 7″ (12.7 × 17.8 cm) prints is $8\frac{1}{2}$″ × 11″ (21.6 × 27.9 cm). This makes it convenient to use margins close to the traditional 3 to 4 to 5 part vertical balance, Figure 40-9. This ratio places the photograph above the mathematical center. It provides for a smaller top margin and a larger bottom margin as compared to the equal size margins. Mounting board size and photograph positioning should be based on good judgment. Photographic contests generally require prints, regardless of size, to be mounted on 16″ × 20″ (40.6 × 50.8 cm) mounting board.

2. Cut the mounting board to size. Calculate the best cut so that the maximum number of boards can be obtained from the larger stock piece.

3. Tack dry mounting tissue to print. Place the print face down on a smooth, clean surface. Position a piece of dry mounting tissue along two edges of the print. Using the warm tacking iron, adhere the dry mounting tissue to the photograph in one small area, Figure 40-10. This will temporarily adhere the two materials together.

4. Trim off excess dry mounting tissue. Turn the print over for trimming, but be very careful not to damage the photograph surface. Use a "razor-sharp" trimmer or an art knife and a steel straight edge. It is important to trim the dry mounting tissue very accurately. Dry mounting tissue extending beyond the edge(s) of a print makes the final mounting very unattractive.

5. Position the print on the mounting board. Use the 3 to 4 to 5 ratio positioning technique or other appropriate method. A T-square and/or measuring rule are necessary for obtaining accurate positioning.

6. Tack dry mounting tissue to the mounting board. Place a piece of clean bond typing paper over the print surface. While holding the print down, lift one corner of the print. Tack the corner of the dry mounting tissue, Figure 40-11. Do the same to the next nearest corner. Never tack more than two corners. This helps prevent wrinkles when the entire surface is heated and pressed.

3 PARTS

4 PARTS

4 PARTS

5 PARTS

(MOUNTING BOARD)

FIGURE 40-9.   The traditional placement of a photograph on mounting board.

7. Make a final position check. Measure the margins and check for squareness. Reposition the print if needed. The tacked dry mounting tissue can easily be lifted without damaging it or the print.

8. Preheat the dry mounting press. Set the temperature according to the instructions with the dry mounting tissue. Make certain the dry mounting tissue is of the right type for the photographs being mounted.

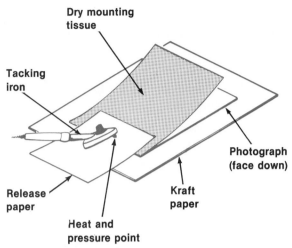

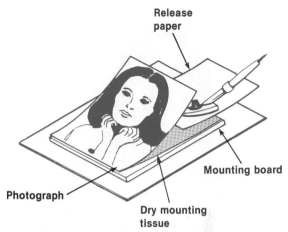

FIGURE 40-10. A warm tacking iron is used to tack the dry mounting tissue to the back of a photographic board. *Seal Products, Inc.*

FIGURE 40-11. Tacking a corner of the dry mounting tissue to the mounting board. *Seal Products, Inc.*

9. Place the several materials in a heated mounting press, Figure 40-12. Use clean bond typing paper or release paper over the print and under the mounting board. This serves to protect the photograph, print-mounting board, and press platens from getting dirty. Leave the top platen up for about 1 minute. This preheats the mounting materials and helps to remove excess moisture.

10. Adhere the photograph and mounting board together. Close the top platen of the mounting press and lock it into place. Set the timer according to the instructions included with the dry mounting tissue.

11. Remove and cool the mounting. Place the mounted photograph, face down, on a clean, smooth surface. Lay a flat metal plate on top of the mounting board. This will keep the mounting flat while it cools.

12. Conduct a quality check. Inspect the mounting to determine if the dry mounting tissue adhered well to both the print and board. If it didn't adhere properly, return it to the mounting press.

FIGURE 40-12. Placing the materials for mounting between the platens of the mounting press.

13. Clean up and enjoy the results. When finished, turn off the tacking iron and mounting press. Clean up and organize the work area. Finally, enjoy and display the finished product, Figure 40-13.

SUMMARY

Mounting photographs is an important part of the entire photographic process. It gives photographers

FIGURE 40-13. The finished mounting is ready for display. *Falcon Safety Products Inc.*

the chance to select their best prints and adhere them to rigid mounting board. When mounted, prints are kept flat and protected from handling damage. They can then be made available for showings to one or many people. There is a wide selection of mounting materials, and several methods and techniques of mounting. Quality results improve the appearance of both black-and-white and color photographic prints.

REVIEW QUESTIONS

Answer these questions to test your knowledge of the unit content.

1. Define "mounting."
2. Which tool is best to use for cutting mounting board?

   A. Box knife   C. Art knife
   B. Scissors   D. Razor blade

3. T/F It is wise to place heavy pencil marks on mounting board when measuring it for size.
4. What is the value of a mat cutter over a box knife?
5. A dry mounting press is designed to provide both _____ and _____.
6. Name the three basic types of adhesive used for mounting photographs.
7. What makes mounting board of "archival" quality?

   A. Thickness and size
   B. Color and finish
   C. Acid-free content
   D. Manufacturing method

8. Which mounting method does not require "visual balance" positioning?

   A. Face   C. Double matting
   B. Window matting   D. Flush

9. When mounting photographs, dry mounting tissue should first be fastened to the "photograph" or the "mounting board?"
10. Why is release paper placed on both sides of the mounting "sandwich" while in the mounting press?

FINISHING AND DISPLAYING PHOTOGRAPHS

OBJECTIVES   *Upon completion of this unit, you will be able to:*
* *Retouch and tone black-and-white prints.*
* *Frame both black-and-white and color prints.*
* *Display photographs in homes, businesses, and for public showings.*
* *Find, enter, and accept results of photo contests.*

KEY TERMS   *The following new terms will be defined in this unit:*
Retouching
Spotting
Toning

INTRODUCTION

Photographs are complete when they are ready to be used. Many photographs are used for printing in publications such as books, magazines, and newspapers. These usually need no further finishing unless they contain blemishes that require retouching. Photographs that will be displayed for direct viewing often will require the mounting and finishing steps presented in this unit and unit 40. Displaying photographs is a "creative art" in itself. This phase of photography should never be taken lightly.

RETOUCHING PRINTS

Unwanted light specks on black-and-white prints often can be covered with special retouching dyes. *Retouching* prints take considerable time and patience. Another term sometimes used for this process is *spotting*. Light spots are caused by opaque dust and dirt specks on negatives. These opaque areas hold back light, thus creating visible white areas on the photographic print, Figure 41-1.

Dark spots on a print are created when tiny transparent areas are contained in the negative. These can sometimes be corrected on the negative by applying special retouching dye. The small size of most negatives, such as 35 mm, make

FIGURE 41-1. Light specks (spots) in a photographic print are often caused by unwanted dust and dirt on the negative.

this difficult for most people, except for the retouching experts, to do. It is possible to scrape dark specks from a print with a sharp tool such as an art knife.

The best defense against spots and specks is cleanliness in the beginning. Keeping negatives, tools, and enlargers perfectly clean often eliminates the need for print retouching. When retouching is needed though, the procedures that follow can be helpful.

1. Obtain the needed tools and supplies. These include one or two small brushes sized #00 or #000, spotting dyes, small dish, art

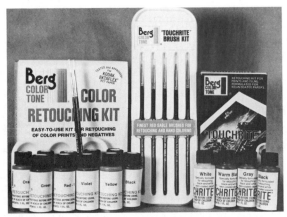

FIGURE 41-2.   Retouching kits are available for both color and black-and-white prints. *Berg Color-Tone, Inc.*

FIGURE 41-3.   A commercial photofinishing station that includes an adjustable, back lighted work easel, flat work surface, and supply storage. *Kreonite, Inc.*

knife, and a supply of distilled water. A small, hand-held magnifying glass, soft tissue paper, and a pair of white cotton gloves are also very useful. Retouching kits include a wide assortment of dyes and brushes, Figure 41-2.

2. Position the print, tools, and supplies. Work on a clean, smooth work surface, Figure 41-3. It is wise to wear white cotton gloves when handling the print. Oils and perspiration from the skin make it difficult for the spotting dyes to adhere.

3. Prepare the spotting dye and brush. Place one or two drops in the small dish. The plastic top of a 35-mm film magazine case works well for this purpose. Place a small amount of dye on the brush tip. Twirl the brush to a sharp point.

4. Apply dye to the print, Figure 41-4. Touch the light speck with the brush tip. Do this several times until the area is completely covered with the dye. Blend the dye with the surrounding print image. The objective is to make the spotting invisible. A quality job has been done when people viewing the finished print never realize that retouching was done.

5. Remove dark spots. Lightly scratch at the emulsion with an art knife, Figure 41-5. Be very careful not to scratch too hard.

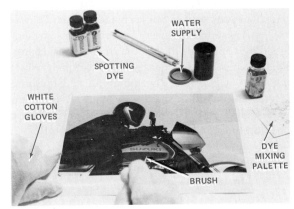

FIGURE 41-4.   Applying the spotting dye to a light spot on a black-and-white print.

Remove enough emulsion to allow the white print paper to show through. Press down on the area with the end of the knife handle. This smooths the emulsion and paper fibers.

6. Some retouching suggestions. Keep the brush as dry as possible. This helps to control the amount of dye being applied to the print. Mix some distilled water with the dye to weaken the color. This helps match the dye color to the print image area. Use the tissue

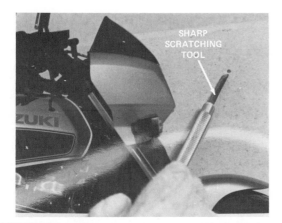

FIGURE 41-5. Removing a dark spot by lightly scratching the print emulsion.

to absorb an excessive amount of dye that may have been applied to the print. Work slowly and deliberately. Use the magnifying glass to make it easier to see. Retouching is a slow process until skills are gained.

TONING PRINTS

Toning black-and-white prints can add an extra effect that attracts attention. It changes an ordinary print into one that creates moods and gives specific impressions of the image. Various toning colors are available, but the most common color is a brownish tone called *sepia*. Other toner colors are blue, green, red, and gold.

Nearly any black-and-white print can be toned. Resin-coated papers sometimes will not work with certain toning solutions. Instructions

included with the toning kit should indicate photographic paper restrictions. Prints are exposed and processed in the usual manner. The most important requirement is for the print to be thoroughly fixed and washed. Some toning chemicals work better when prints are fixed in a solution without hardener. Processing chemicals left in a print will interfere with the toning chemistry. This often creates blemishes that severely reduce the overall quality of the toned print.

1. Obtain a sepia toning kit. A kit contains a bleaching solution and a sulfide darkening solution. Sulfide is a compound of sulfur and oxide or ether. It should be used in a well-ventilated area.

2. Prepare for toning. Three print processing trays will be needed, Figure 41-6. Pour some bleaching solution in one tray. Use the second tray for wash water by attaching a tray wash siphon. The third tray will contain the sulfide darkening solution. Place the trays in a standard print processing sink. In that way, spills will not be damaging to tables and counter tops.

3. Bleach the print. Lift one side of the tray to move the solution to the other side. Insert the print in the same manner as used for processing. Rock the tray gently. Watch the print until the image changes to a yellow-brown color. At this point, all visible black metallic silver will be removed. This usually takes 1 minute or less.

FIGURE 41-6. The typical tray setup for sepia toning a black-and-white print.

4. Wash the print. Place the print in the wash tray. Be sure to use a print tong to transfer the print from the bleach tray. Bleach solution can be harmful to skin and especially to open sores. RC paper should be washed for 2 minutes and fiber-based paper for 5 minutes.

5. Tone the print. Immerse the washed print in the sulfide darkening solution. Continuously rock the tray gently until the image will turn no darker. The time period depends on the strength of the toning solution and the content of the image area.

6. Wash and dry the print. Wash the print for 10 minutes, and dry in the normal manner. The print has now taken on a "mood" change due to its changed color.

7. Clean up the work area. The bleach solution can be saved and reused for additional prints. Place it in a clearly labeled container that can be securely capped. The toning solution cannot be reused; thus, it must be discarded. Pour it down a drain especially designed for chemical solutions. The strong odor of the toning solution will linger for days in a regular drain.

FRAMING PHOTOGRAPHS

Frames add to the beauty and practicality of a photograph, Figure 41-7. Selecting the "right" frame enhances a photograph by drawing attention to it.

FIGURE 41-7. Picture frames help display photographs for viewing enjoyment. *The Holson Company*

Frames also make it possible to display photographs in a wide variety of locations. Walls and furniture are the two main locations for framed photographs.

Frames are made of wood, metal, and plastic, They can be premade, precut, or self-made. Premade frames can be purchased in a variety of retail stores. It takes only a few minutes to insert a photograph and be ready to display the finished product. Precut frames take longer to prepare but permit a greater variety of sizes and shapes. Many precut frames are made of metal (aluminum). These are fastened together with metal brackets and screws. They help make homes and offices pleasant and attractive.

Self-made frames can be constructed from many kinds of wood. It takes special skills to operate the needed equipment. Any photographer possessing this extra talent will gain additional pleasure in designing, constructing, and finishing frames for the waiting photographic prints.

DISPLAY METHODS

Every effort should be made to place photographs where they can be seen. One of the best ways to display snapshots is to place them in photo albums, Figures 41-8 and 41-9. Commercial display packaging of this type permits convenient storage, filing, and viewing by people of all ages. The photo albums shown in Figure 41-9 are designed to appear as book volumes. Labels can be applied to the front of the albums, making them easy to identify.

Slides also must be made convenient for viewing both on an individual basis and for large groups. Back lighted slide sorters are useful for organizing slides, Figure 41-10. These units contain a white plastic angled front and have built-in channels so several slides can be set in place. A lamp under the channeled plastic illuminates the slides. Projection equipment is needed for showing 35-mm slides to larger groups, Figure 41-11. Audio scripts often are added to a series of slides to make them more meaningful.

Mounted prints are easily displayed using gallery wall brackets, Figure 41-12. The brackets are fastened to a wall with self-adhesive tape or screws.

FIGURE 41-8. Bi-directional pocket pages in these photo albums allow ease of use for the photographer. *The Holson Company*

FIGURE 41-10. A slide sorter makes it convenient to organize a series of 2 × 2 slides. *University Products, Inc.*

FIGURE 41-9. Miniature photo albums (books) with large capacity are very popular for viewing and storing photographs. *The Holson Company*

FIGURE 41-11. A projector and audio system are useful when showing a slide-audio series to a group of people. *Creatron Inc.*

The mounted prints are then placed in the slot where friction rollers gently grip the mounting. This and other system designs permit photographs to be displayed at various public showings. It is important to use attractive picture holders that permit convenient installation and take-down of the mounted prints.

PHOTO CONTESTS

Photo contests can be exciting, challenging, and educational. A competitive event often provides the reason to "plan, take, process, and finish" one or more photographs. Some people believe

FIGURE 41-12.   Special gallery wall brackets make it convenient to display mounted photographs.

contests provide the ultimate opportunity to display photographs.

Locating and entering a contest can be time-consuming but very interesting. Companies and organizations are primary sponsors of competitive events. They are often involved in such a program to advertise and promote themselves. This is how they can justify the associated expenses. Photo contests are advertised in newspapers, magazines, direct-mail, printed literature placed in photo retail stores, radio, and television.

Read the contest rules very closely. Check eligibility requirements for areas such as age range, for amateur or professional photographers, kind of entries requested, number of entries permissible, and the subject matter. Also, should the photograph have been published previously or must it be new work. Often a photographer must give up all future rights of the photograph to the sponsor. Beware of contests that require large entry fees. Any photo contest having a large entry fee indicates that the sponsor is interested in making a large amount of money.

Finally, review the list of prizes. Make certain that the prizes are worth the time and trouble.

The best prize is "feedback" from the judges. A critique from a complete stranger can be very valuable. Accepting the challenge of a photo contest means opportunities for learning and being rewarded.

SUMMARY

Photographs can be retouched, toned, framed, and displayed. They also can be entered into contests that provide opportunities for learning and recognition. Retouching should be kept to a minimum. Every effort needs to be made to keep dust and dirt from equipment and film. The major cause of specks and spots on photographs is the lack of cleanliness. Toning adds a dimension of color that makes photographs stand out from the crowd. Proper framing and display give photographs the final touch. Remembering that photographs are made to be seen should give sufficient cause to finish them in the best manner possible.

REVIEW QUESTIONS

Answer these questions to test your knowledge of the unit content.

1. T/F Light specks and dark spots on a photographic print are created by the same specific problem.
2. What material is used to cover unwanted light specks on a print?

   A. Toning dye   C. Photo paint
   B. Retouching dye   D. Dark ink

3. T/F Cleanliness from beginning to end is the best defense against specks and spots.
4. White _____ _____ should be used when handling and retouching prints.
5. T/F The best retouching job is one that is not noticed by a viewer.
6. Which print toning color is probably the most popular?

   A. Blue   C. Red
   B. Green   D. Sepia

7. How many solutions are used for toning a print in addition to water?

A. 1   C. 3
B. 2   D. 4

8. What is the purpose of framing a photographic print?

9. T/F Slides (2″ × 2″ size) and prints are equally convenient for showing to both small and large groups.

10. The best prize in a photo contest is _____ because it provides an opportunity for additional learning.

UNIT 42

PRESERVING PHOTOGRAPHS

OBJECTIVES   *Upon completion of this unit, you will be able to:*
* *Discuss the causes of photographic image deterioration.*
* *Display photographs in a manner that will help preserve them.*
* *Rephotograph prints to reduce or eliminate unattractive stains.*

KEY TERMS   *The following new terms will be defined in this unit:*

Archival   *Image Deterioration*
Dark Storage   *Relative Humidity*

INTRODUCTION

Photographs serve as visual records of people, places, and events. They provide evidence of what took place in past generations. Much can be learned from studying silver-based, black-and-white photographs of yesteryear. The same is true of dye-based color prints and transparencies. Future generations will be interested in learning about current happenings. This information can be made available if effort is made to prepare photographic images for a long life.

CAUSES OF IMAGE DETERIORATION

There are three major causes of *image deterioration*. These are processing, storing, and handling, Figure 42-1. Each of these is important as single problems, but all three must receive equal attention. Weak treatment of just one part of a major cause can cause photographic image deterioration.

PROCESSING   It is critical to follow film and print processing instructions closely. Manufacturers continually conduct research directed at the *archival* needs of photographic products. The cause of film and print deterioration is the chemicals left after

FIGURE 42-1. Old photographs often acquire a deteriorated appearance after years of storage and handling.

processing has been completed. Unused silver compounds break down and cause a yellowish stain to appear on black-and-white negatives and prints. Fixer that is not washed completely from both negatives and prints will eventually cause a discoloration.

STORING Negatives and prints are badly affected by improper storage. Temperature and humidity are the major factors that must be considered. **Heat** is always hard on photographic materials. It is wise not to allow the temperature to go higher than 70 °F (21 °C). Colder storage temperatures are better. Temperatures colder than 45 °F (7.2 °C) and even below freezing increase storage life from a few years to 200 years and more.

Humidity must be tightly controlled. High relative humidity is very damaging to negatives and prints. It is best to keep the level of relative humidity below 50%. Whenever possible, 30% to 40% relative humidity should be maintained in any area used for photographic storage.

Light can cause photographic materials to decompose rapidly. "Dark storage" is commonly used by collectors and libraries for negatives, prints, and slides. This is an area that can be kept dark most of the time. Light needed to see in the storage area for adding or removing materials should be free of ultraviolet rays. Sunlight and fluorescent light contain high amounts of ultraviolet rays. The best light to use for short periods of time is incandescent (standard electric light bulbs).

For long storage, it is useful to place processed slides in special storage envelopes or sleeves, Figure 42-2. These sleeves are made of materials that resist heat, humidity, gases, and light. After the slides are loaded, the sleeves can be kept in a refrigerator, heat-resistant file cabinet, or in an office or home area that can be kept cool and dry.

HANDLING Many photographs and slides are damaged while being viewed. Light used to show slides with a slide projector will remove a slight amount of colored dye each time. After a large number of showings, there is a significant color change. The best way to guard against this problem is to make duplicate slides for use in the projector.

Keep the original slides in a proper storage facility.

Touching photographic prints while viewing can be very harmful, Figure 42-3. Oil and dirt from the skin are harmful to film and print emulsions. One of the best ways to protect prints is to mount and frame them. In this way, people who do not understand the problems associated with touching photographs will either handle the mounting board or the frame.

FIGURE 42-2. For long slide life, sleeves designed to hold 20, 2″ × 2″ slides should be made of archival materials. *Print File, Inc.*

FIGURE 42-3. Improper handling and touching of unprotected photographs leave a deposit of skin oil and dirt on the image emulsion.

PRESERVING WHILE DISPLAYING

Attention to light, temperature, and humidity is critical for long periods of display. This is especially true for color photographs but also should be considered for black-and-white prints. The following pointers should be observed whenever possible.

1. Keep all photographic prints out of direct sunlight as much as possible.

2. Use tungsten light where possible and practical for viewing and extended periods of display.

3. Hang photographs in cooler areas of a home and office.

4. Observe humidity levels and make every effort to keep the relative humidity below 50%.

5. Obtain two prints of a photograph from a professional studio—use one for display and place the other in dark storage.

6. Use ultraviolet absorbing materials to reduce the level of UV light reaching photographs. UV shields can be placed over fluorescent tubes and UV plastic sheets can be used in picture frames.

7. Think before selecting a location to display one or more photographs. Photographs are prized possessions that need protection and care.

REPHOTOGRAPHING PRINTS

Photographs sometimes obtain stains or marks that reduce their appearance. This is especially true of black-and-white photographs that are 20 or more years old. Chemical treatments can be applied to the stained area of a print. This is usually successful only to expert photo restoration personnel. Amateurs find this technique somewhat difficult.

One of the best methods to reduce or eliminate a print stain is to rephotograph the print. The basic principle involves using a high quality copy film and a filter of the same color as the stain. For example, a yellow colored stain can be reduced or eliminated by photographing through a yellow filter. The filter absorbs its own color. A typical arrangement includes using a standard copy stand, a 35-mm camera, a bellows or extension tubes, and a selection of filters, Figure 42-4.

SUMMARY

The concern about preserving photographs is becoming more important. Each year there are more photographs and processed film that need preserving. The problems associated with photographic preservation are many. The main concerns revolve around heat, humidity, and light. Controlling these three factors helps make photo materials last for many years. New photographic print papers and film emulsions raise the problem of longevity. Estimations and preaging techniques can be applied, but only years of time will provide an accurate answer.

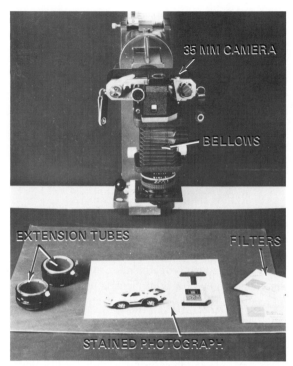

FIGURE 42-4. Rephotographing stained prints to reduce or eliminate visible stains can be done with a copy stand and associated equipment.

Much research remains to be done in this important area of photography. There are, though, many products available now that are designed to help protect and preserve photographic images, Figure 42-5.

REVIEW QUESTIONS

Answer these questions to test your knowledge of the unit content.

1. T/F Both black-and-white and color photographic prints deteriorate with age.
2. What processing problem causes film and prints to deteriorate?
3. Unused silver compounds left on processed black-and-white film often will cause what stain color to appear?

   A. Brown   C. Yellow
   B. Green   D. Black

4. Temperature and _____ are two of the most serious causes of photographic image deterioration.
5. T/F Colder temperatures will help preserve negatives but not photographic prints.
6. The most appropriate relative humidity to maintain in photographic storage areas is:

   A. 30%   C. 70%
   B. 50%   D. 90%

7. Which kind of light causes photographic materials to decompose most rapidly?

   A. Fluorescent   C. Tungsten
   B. Mercury vapor   D. Sunlight

8. What is a good way to reduce photograph damage from viewing and handling?
9. List the three main decomposition problems associated with displaying photographs.
10. T/F Rephotographing a stained print requires the use of a special manufactured camera.

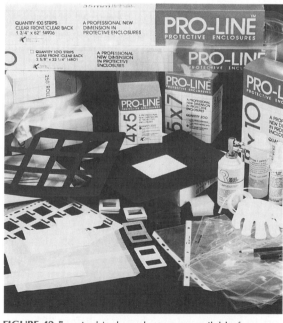

FIGURE 42-5. Archival products are available for negatives, prints, and slides. *Kleer Vu Plastics Corporation*

CHAPTER NINE

PHOTOGRAPHY: NOW AND THE FUTURE

INSTANT PHOTOGRAPHY

OBJECTIVES   *Upon completion of this unit, you will be able to:*
- *Discuss the basics of instant cameras.*
- *Describe instant film products.*
- *Use an instant camera and make acceptable prints.*
- *Summarize the history of instant photography.*

KEY TERMS   *The following new terms will be defined in this unit:*

Dry Instant Film   *LED Peel-apart Instant Film*

Instant Photography   *Sonar Focusing*

INTRODUCTION

Instant photography is both easy and enjoyable. It gives people the opportunity to squeeze the shutter release and within a short time see the image on a print. Extra exposures can be taken immediately if additional prints are wanted. Also, it is possible to correct mistakes right away. Amateur photographers make the greatest use of instant photography for taking snapshots of family and friends, Figure 43-1. Serious amateur and professional photographers make good use of instant cameras and film. They take test exposures and make useful prints for their record files.

INSTANT CAMERAS

A variety of instant cameras have been made available in recent years. All of them are easy to use for people of various ages and photographic abilities. A wide series of cameras is generally available from the manufacturer, Figure 43-2. Exposure with nearly all instant cameras is automatically controlled. Some cameras simply contain more features; thus, the purchase cost is increased. New models of instant cameras continue to be released on a regular schedule.

Instant cameras have the same basic parts as conventional cameras. These include the following.

LENS. Some instant cameras have fixed-focus lenses that permit pictures to be taken as close as 7' (2.1 m) through infinity. Other cameras have adjustable lenses permitting operator adjustments. A few of the "professional model" instant cameras contain interchangeable lenses. The number of elements in a given lens varies with the quality level of the total camera.

FIGURE 43-1. Instant cameras give amateur photographers the freedom to take and quickly share their photographic results. *Polaroid Corporation*

FIGURE 43-2. A series of instant cameras gives photographers the choice of selecting a camera to meet their needs. *Polaroid Corporation*

FIGURE 43-3. An instant camera that uses Sonar sound waves to automatically focus the lens. *Polaroid Corporation*

APERTURE. Apertures are fixed on the economical model instant cameras. Sometimes there is an operator choice of two apertures. Weather symbols marked near the adjusting lever tell the operator the best choice. The programmed instant cameras automatically select the correct aperture for the intended picture scene.

SHUTTER. Speeds can be varied to obtain the correct exposure. Top shutter speeds on many instant cameras are 1/125 to 1/500. Some of the expensive cameras permit time exposures as long as 14 seconds. Automatic instant cameras permit some operator control by moving a light-dark image control lever. This changes the shutter speed and leaves the aperture constant.

VIEWFINDER. The vast majority of instant cameras are of the viewfinder camera style. This makes the viewfinder separate from the lens. Some of the instant models are of the single-lens reflex (SLR) design. This, of course, permits viewing the image through-the-lens.

FLASH. Nearly all models of instant cameras contain built-in electronic flashes. A sensor system automatically determines the need for flash lighting.

Some economical models use miniature flash bulbs made into a horizontal bar (see Figure 43-2). Batteries for the electronic flash and sensor systems are either loaded into the camera separately or included with each film package.

Automatic focusing has been designed into some top-of-the-line models, Figure 43-3. An autofocus transducer near the lens of the camera makes readings by using sound waves, Figure 43-4. The *sonar* rangefinding system automatically focuses the camera lens in any kind of light from two feet (0.69 m) to infinity. The camera emits sound waves that strike the subject. They bounce back in milliseconds, thus informing the Sonar system of the distance between the camera and the subject. The lens is quickly and accurately focused by a microprocessor contained within the camera.

INSTANT FILM

Each instant camera is designed to accept one type of instant film. Once a camera is selected, the only choice often available to the photographer is

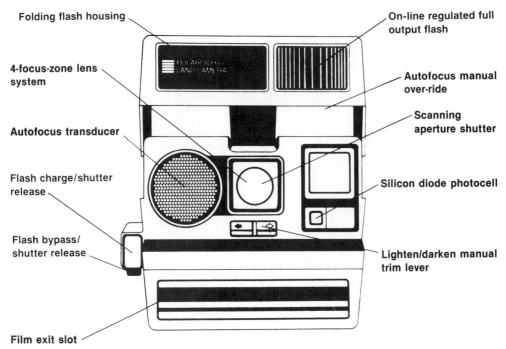

Folding flash housing

4-focus-zone lens system

Autofocus transducer

Flash charge/shutter release

Flash bypass/ shutter release

Film exit slot

On-line regulated full output flash

Autofocus manual over-ride

Scanning aperture shutter

Silicon diode photocell

Lighten/darken manual trim lever

FIGURE 43-4. The basic parts of a standard instant camera. *Polaroid Corporation*

FIGURE 43-5. A wide variety of instant film products is available to amateur and professional photographers. *Polaroid Corporation*

whether the prints should be color or black-and-white. There are a wide variety of instant films available for both amateur and professional photographers, Figure 43-5. The major markets for most instant film products involve amateur photographers, Figure 43-6. The snapshot-type prints are handled and placed into photo albums just like conventional prints. Professional photographers utilize instant films for proofing. These films help confirm proper exposure and composition before taking pictures with conventional film.

There are two basic types of instant film products. The oldest and original type is called *peel-apart*. This type involves pulling the exposed print from the camera. Developing chemicals within the film pack are distributed over the two layers of film at this point. After about 1 minute, the print and negative are peeled apart. The negative layer is discarded, and the print must receive a coat of protective lacquer. This is a messy process and has been replaced by a newer, more convenient type.

The second type of instant film is known as *dry film.* The processing chemicals are sealed within the layers of the print material, Figure 43-7. After the image is exposed, the print packet is ejected automatically from the camera. The processing fluid

FIGURE 43-6. Amateur photographers take most instant prints of family, friends, and surroundings. *Polaroid Corporation*

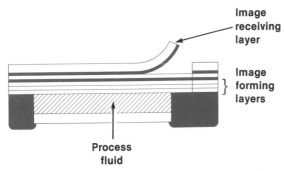

FIGURE 43-7. The process fluid is sealed within each film packet of the "dry" type of instant film.

is spread over the positive and negative materials to begin development. The processing is a combination developer and fixer solution that makes the image visible and permanent. The image begins developing immediately after leaving the camera. Full development is generally completed in 2 to 3 minutes.

Instant slide film makes it possible to use a standard 35-mm SLR camera to expose the special film. Immediately after exposure, the film can be processed by using a power processing unit, Figure 43-8. A slide mounting unit permits quick preparation of the finished slides, Figure 43-9. Three 35-mm films are available: color, black-and-white continuous-tone, and black-and-white line copy film. This type of system is useful for preparing slides to be used for illustrated presentations. Also, slides can be projected directly into hanging picture frames for customers to immediately view their portraits. The color quality does not match standard emulsion slide films, but the results are acceptable for specific uses.

Instant 8″ × 10″ (20 × 24 cm) prints and overhead transparencies can be made from 35-mm slides, Figure 43-10. No darkroom or wet chemicals are needed to produce these finished products. Many other instant accessories and materials are available to amateur and professional photographers, Figure 43-12. Several different types of

FIGURE 43-8. A power processor can be used to process 35-mm instant film. *Polaroid Corporation*

instant film adaptors have been designed for SLR and view cameras. This makes it possible to use cameras familiar to the photographer and have the advantage of instant (within 1 to 3 minutes) results.

FIGURE 43-9. Instant 35-mm slide film can be quickly processed and mounted into frames. *Polaroid Corporation*

FIGURE 43-10. Instant prints and overhead transparencies can be made from 35-mm slides with this system. *Polaroid Corporation*

TAKING INSTANT PICTURES

There is little difference in taking instant pictures and conventional pictures. The photographer should use good composition guidelines and remember the limitations of the camera being used. The following steps will help when using an instant camera:

1. Prepare the camera. Remove camera from the protective case and unfold it if necessary.

2. Load the camera with film. Select the correct color or black-and-white film for the camera. Each film pack generally contains 10 exposures.

3. Make needed manual adjustments. Few manual settings can be made on instant cameras. Most adjustments are made automatically by integrated circuits (microchips) and special detectors, Figure 43-11.

4. Take the picture. Aim the camera and squeeze the shutter. Hold the camera securely with both hands. Make certain nothing obstructs the lens, such as having a finger partially covering its field of view.

5. Remove the print. Immediately after the exposure, the print will be ejected from the camera. Place it on a flat surface and watch it develop. This can take place in a lighted area.

6. Evaluate the results. If the print is too light or too dark, make the needed adjustment and take another exposure. When the film is cold, it reacts more slowly to light than when warm.

7. Enjoy the results, Figure 43-13. Instant prints provide quick feedback for the photographer and the subject as well as other people interested in the photo scene.

FIGURE 43-11. Features of this instant camera include computer chip controls, timed exposure control, interval timer for taking a series of pictures, multiple self-timer for producing ten copies of the same picture, and a multiple option that permits the photographer to make five exposures in the same picture. *Polaroid Corporation*

HISTORICAL HIGHLIGHTS

The first camera to produce finished photographic prints was demonstrated before the Optical Society of America in the spring of 1947. The person responsible for this advancement in photographic science was Dr. Edwin H. Land of the Polaroid Corporation. He initiated his plan to design a one-step process for making pictures in 1943. His goal was to produce a camera that would permit the photographer to concentrate on the picture rather than on the process.

The first instant camera was placed on the market for public sale on November 26, 1948 in Boston, Massachusetts. It was called the Polaroid Land Camera, Model 95. Advancements came rapidly in both cameras and film. In the first 30 years,

over 80,000,000 Polaroid Land Cameras were sold.

Other companies tried to produce and market instant cameras and film products. Most were either totally unsuccessful or managed a short stay in the market place. Today, The Polaroid Corporation manufactures and distributes instant cameras, film products, and accessories for markets throughout United States and the world.

SUMMARY

Instant photography is an important part of the total photographic industry. Technically, this process is not instant, but the finished print or slide is produced within 1 to 3 minutes after squeezing the shutter. That makes the process fast enough to fully qualify to name it "instant" photography.

FIGURE 43-12. Instant color photographic images can be pressed onto non-photographic material to make "image transfers," *Polaroid Corporation*

Instant photography is primarily supported by the amateur market called snapshot photography. It is also widely used by professional personnel. Instant pictures are taken of portrait scenes, product setups, and landscape scenes so that composition and lighting can be evaluated. Afterward, SLR and large-format view cameras are used to capture the image on conventional film products. Many times, though, the professional photographer will use the instant print or slide as the final result. Continued improvements in this aspect of photography will provide a continued challenge to "standard processing" continuous-tone photography.

REVIEW QUESTIONS

Answer these questions to test your knowledge of the unit content.

FIGURE 43-13. Enjoying the results of an "instant" print. *Polaroid Corporation*

1. T/F Instant photography permits results to be seen beginning as quickly as seconds after squeezing the shutter release.
2. T/F In general, instant cameras are more difficult to operate than 35-mm cameras.
3. Many models of instant cameras contain a _____ _____ lens which allows pictures to be taken from a short distance to infinity.
4. Maximum shutter speeds on some instant cameras may reach:

    A. 1/100   C. 1/500
    B. 1/200   D. 1/1000
5. Most instant cameras are classified into which camera category?

    A. Viewfinder   C. View
    B. SLR   D. TLR
6. T/F Sonar sound waves are used on some instant cameras to automatically focus the lens.

7. Name the one major manufacturer of instant cameras and film products.

8. Which type of instant film is in common use today?

9. T/F Instant film can be exposed in conventional cameras, such as a 35-mm SLR.

10. Who is credited with inventing the first instant camera?

————— UNIT 44 —————
ELECTRONIC PHOTOGRAPHY

OBJECTIVES *Upon completion of this unit, you will be able to:*
- *Describe the two categories of video imaging.*
- *Discuss the principles of photo-video imaging.*
- *Explain the operational system for Photo CD's.*
- *Summarize how computers, electronics, and photography have been combined.*

KEY TERMS *The following new terms will be defined in this unit:*

Electronic Imaging	*Optical Disc*	*Pixel*
Fiber Optics	*Photo CD*	*Streak Camera*

INTRODUCTION

There have been many advances in electronic photography in recent years. It is now possible to capture both moving and still images in electronic memory that look very acceptable to the human eye. Television serves as strong proof. The image resolution (quality) of television cannot currently match that of film. All pictures, electronic or photographic film, contain minute picture elements. These are called *pixels.* There are about 10 million pixels in one 35-mm frame of quality color film. The state of the art with standard television gives a resolution of about 250,000 pixels in the same space. Progress in electronic imaging that will be suitable in cost for the average consumer is being made rapidly. An example of this is with high definition television (HDTV).

VIDEO IMAGING

Video cameras and equipment have been available for several years. It is a common sight to see news and sports reporters interviewing guests, Figure 44-1. It has become so common that most people pay little attention to the equipment or the repor-

ter. Much of the time people have great interest in being "part of the picture" because they can be seen on the television screen in homes, offices, and businesses. The visual image even when temporary is very important to people.

Two general types of complete video imaging are in use. These are moving and still video imaging.

MOVING VIDEO IMAGING. This type is designed to capture moving images on magnetic tape for the purpose of showing on a television screen, Figure 44-2. The image resolution is less than a photographic print, but the advantage is instant storage and retrieval. The camera equipment shown in Figure 44-2 is designed for general purpose imaging. It is convenient to use and has many adjustments and much flexibility.

Electronic imaging is finding valuable use in motion analysis systems, Figure 44-3. A typical system includes a high-resolution camera, a magnetic tape recorder, a high-resolution monitor, and a sophisticated control panel. These systems can record up to 1000 full-frame images per second. Partial format pictures can be recorded at 6,000 pictures per second. Instant replay in variable slow

FIGURE 44-3. A high-resolution electronic camera, receiver, and control system. *Eastman Kodak Company, Motion Analysis Systems Division*

FIGURE 44-1. A television studio provides an excellent environment to electronically record both visual images and audio sounds. *JVC Company of America*

FIGURE 44-2. A video camera, often called a camcorder because it is a combination camera and recording unit, is convenient to use after minimal instruction. *JVC Company of America*

FIGURE 44-4. A high speed motion analysis system permits close-up study of moving objects that are captured with an electronic motion camera. *Eastman Kodak Company, Motion Analysis Systems Division*

FIGURE 44-5. An SLR still video camera, interchangeable lenses, and magnetic image disks. *Electronic Photography and Publishing Division, Sony Business and Professional Group*

motion makes it possible to visually study the image, Figure 44-4. Images can be increased or decreased in size from the control panel.

STILL VIDEO IMAGING. In August of 1981, the first still video camera was announced, Figure 44-5. The camera with its interchangeable lenses looked much like a standard 35-mm camera. The major difference was that it used no photographic film. Instead, a *photomagnetic disk* located at the camera's image plane was used to record the still images. Current electronic camera systems transmit the individual images directly to digital storage units, Figure 44-6. Portrait photographers enjoy the combination of film and electronic camera-viewing systems, Figure 44-7. From electronic systems, full color

FIGURE 44-6. A digital still camera system that is capable of high quality imagery in both black-and-white and color. *Eastman Kodak Company*

FIGURE 44-7. An electronic camera and instant replay system that includes complete computer control for color balance, cropping, and sizing of the finished image. *Electronic Photography and Publishing Division, Sony Business and Professional Group*

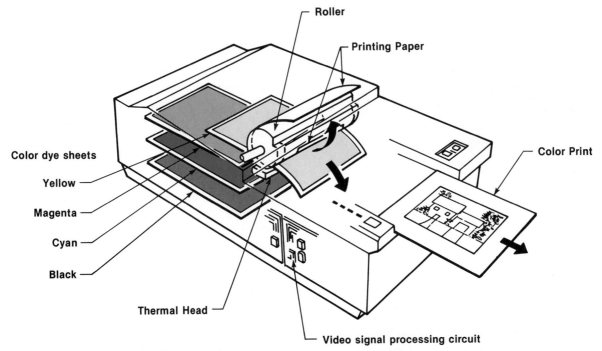

Roller

Printing Paper

Color dye sheets

Yellow

Magenta

Cyan

Black

Thermal Head

Video signal processing circuit

Color Print

FIGURE 44-8. Images stored in electronic memory can be made into full color prints with thermal printers. *Electronic Photography and Publishing Division, Sony Business and Professional Group*

prints can be made on thermal (heat) printers, Figure 44-8. The prints are produced by a thermal printing head that controls the amount of yellow, magenta, cyan, and black dyes.

Still video imaging has many applications, Figure 44-9. Electronic signals are easily transmitted over telephone wires from one location to another. Home, business, and industrial needs will provide many applications for still electronic photography. Immediate viewing of the recorded still images on television sets is a strong asset of a system of this type. Other advantages include no deterioration of picture color quality and quick access to any of the pictures.

PHOTO-VIDEO IMAGING

Electronics and photography are combined in cameras used for research. One group of these imaging devices is called *streak cameras*, Figure 44-10. These high-speed image converter cameras are used to record events that cannot be accomplished with standard film cameras. The lens is coupled with a television-like image tube. It produces the image with an electron beam on the within-camera screen, Figure 44-11. This permits the operator to see the image in the viewfinder.

The selected images are recorded on highly sensitive film of either the standard or instant type. The image is transmitted directly to the film through *fiber optics.* These are very thin transparent fibers of glass or plastic enclosed in special material. These fibers transmit light throughout the entire length by internal reflections. It is convenient to transmit light in this manner because the fibers can be easily curved and bent to reach selected parts of the camera. A typical use of this camera is to record events that take place very rapidly, Figure 44-12.

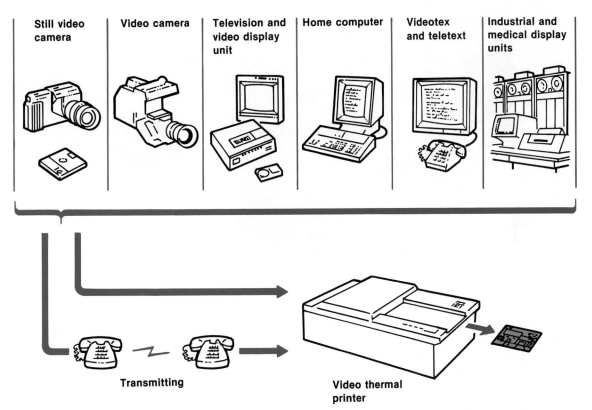

FIGURE 44-9. An example of the many uses of still video photography. *Electronic Photography and Publishing Division, Sony Business and Professional Group*

FIGURE 44-10. A streak or image converter camera converts photons of light into electrons that are used to expose high-speed film. *Hadland Photonics, Ltd.*

Streak cameras can expose film up to 1 million frames per second.

A second type of photo-video imaging is designed for the everyday amateur photographer. One such system is called, "Photo CD." It combines standard 35mm silver-based photography and electronic-based photography by using compact disc, commonly called CD's, Figure 44-13. The original series of pictures is taken with a standard 35mm viewfinder or single lens reflex camera. Processing is completed in the normal manner to provide color negative film and regular sized color prints or color slides. From here, the silver-based negative and slide images are entered into the "Photo CD" system.

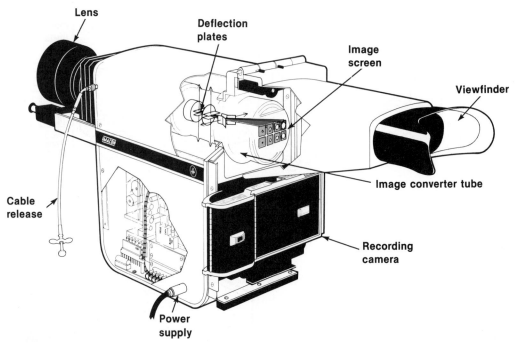

FIGURE 44-11. A partial inside look at the image converter camera shown in Figure 44-10. *Hadland Photonics Ltd.*

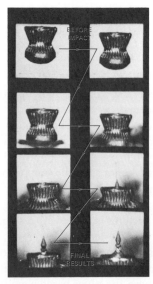

FIGURE 44-12. This series of frames shows what happens when an airgun pellet hits a steel plate. *Hadland Photonics, Ltd.*

RECORDING A PHOTO CD

To transfer color images into the photo CD system, the photographer must take the processed color negative or slide film to an authorized photofinisher or photography dealer. Using special Photo CD transfer equipment, an optical compact disc of the photographic images is created. The Photo CD Transfer Station consists of five major components, Figure 44-14. These are a high resolution film scanner, a computer workstation, image processing software, a photo CD writer (recorder), and a thermal printer.

The images stored on the optical photographic compact discs are complete digital records of the photographic images. The electronic images actually contain more digital information than current television and computer screens can display. It has been stated that the image quality of compact discs is four times greater than that of high definition television. The standard "Photo CD" can be

Standard Photographic Process

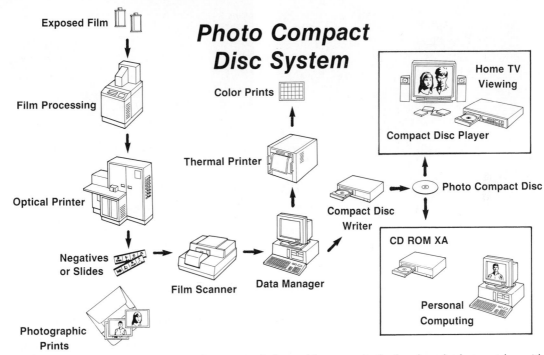

FIGURE 44-13. The "Photo CD System" (compact disc) combines standard silver-based photography with electronic-storage and retrieval associated with computer technology. *Eastman Kodak Company*

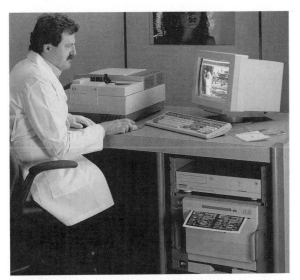

FIGURE 44-14. A photo CD Transfer Station that is used to transfer silver-based film images to electronic images on compact discs. *Eastman Kodak Company*

used to store as many as 100, 35mm frames of information.

USING A PHOTO CD

During the process of making a photo compact disc, an index print is made for the front cover of the compact disc case, Figure 44-15. The index print is made with the thermal printer which is controlled with the data manager computer and software (See Figure 44-13.) The numbers identified with each picture permits convenient and quick retrieval of each image on a regular television receiver or on a personal computer.

A photo compact disc serves as an excellent organizer for photographic prints and slides. It is possible to combine images from color negative film and slides on the same photo compact disc. Once

FIGURE 44-15. Management of the photographs on the compact disc is made easy through the numbered small index prints. *Eastman Kodak Company*

FIGURE 44-16. With a CD ROM XA input unit, it is possible to view and utilize photographic compact disc images on personal sized computers. *Eastman Kodak Company*

the disc has been prepared, it is possible for the user to view the images conveniently on a home television set using a photo compact disc player. The system permits the viewer to enlarge a portion of the original photograph for closer viewing. Also, if there are photographs that should not be shown, they can easily be bypassed. Images can be cropped and rotated from horizontal to vertical or vice versa.

Using personal computers and appropriate software, the electronic photographs can be incorporated into desktop publishing systems, Figure 44-16. Other uses include multimedia presentations, document management for such things as insurance company accident photographs, real estate company photographs of homes and commercial buildings offered for sale, and numerous medical and governmental applications.

SUMMARY

Electronic photography is here to stay. Several uses have already been made for this instant system of recording visual images using electronic signals. The image resolution is somewhat less than silver-based film, but certain applications give this medium an excellent future. Combining silver-based imaging with electronic imaging has proven to be successful. In this way, the qualities of each provide a hybrid system that allows flexibility and convenience for all people who use visual imaging. Electronics and photography have the potential of benefiting everyone.

REVIEW QUESTIONS

Answer these questions to test your knowledge of the unit content.

1. About how many pixels are contained in a 35-mm frame of quality film?

A. 1000   C. 100,000

B. 10,000   D. 10,000,000

2. T/F Electronic imaging quality is equal to silver-based film.

3. Moving video imaging systems are normally shortened to the name _____.

4. T/F Electronic imaging systems permit images to be recorded at very high speeds.

5. Photo-electronic digital cameras look and work much like the standard _____ 35-mm cameras.

6. T/F Electronic camera images are often linked directly to computers for full image control.

7. Streak cameras can be used to record how many image frames per second?

A. 100,000   C. 1000

B. 1,000,000   D. 10,000,000

8. Electronic _____ and framing are possible with video-display systems.

9. T/F Silver based photography is a critical component of photographic compact disc technology.

10. What are three advantages of recording and using Photo CD's?

GLOSSARY

Absorb: To take in or consume, such as a filter absorbs certain colors of light.

Accelerator: One of the four subparts of a developer solution; purpose is to increase the activity of the reducing agents.

Activator Solution: An alkaline solution used to develop stabilization papers containing the developing agent in their emulsions.

Acutance: Describes how photographic film and paper are able to record image sharpness.

Additive Colors: The primary light colors of red, green, and blue. Each color contains about 1/3 of the wavelengths in the visible spectrum. Combining all three additive colors produces white.

Adjustable Camera: A camera within any of the four general categories that permits operator adjustment of f-stops, shutter speeds, and lens focus.

AF: Letters that serve as an abbreviation for "automatic focus" or "autofocus." Many contemporary cameras have lens systems that are of the auto-focus type.

Agitate: To provide motion in an established pattern to circulate chemistry over film surfaces during processing.

Airbells: Bubbles of air that adhere to the surface of film. This causes clear or near-clear spots on the processed film because developer could not reach the film.

Analyzer: An electronic instrument used to measure color printing images. It helps to select the correct filtration when making color prints.

Angle of View: The amount of coverage of a given camera lens or hand-held light meter.

Aperture: The opening through which light enters a lens.

Aperture Priority: A camera feature that allows the photographer to manually set the aperture, and then the camera-computer selects the appropriate shutter speed.

Aperture Ring: The part of the lens barrel that can be turned to adjust the iris diaphragm to different f-stops.

Archival: Relates to the preservation of photographic images on film and paper for a long period of time.

Artificial Light: Light created by any source other than the sun.

ASA: The now outdated speed-rating system for photographic film. The initials did stand for American National Standards Institute. See ISO.

Auxiliary Close-up Lens: A lens that can be placed into position over the regular camera lens to provide close-up pictures.

Available Light: Refers to the normal lighting found inside a building.

Background Light: The light used to illuminate the physical background behind the portrait subject.

Backgrounds: A visual image, generally behind the subject, that creates a neutral or colorful background for portrait photographs.

Baffle: A device used to regulate or block the continuation or movement of something, such as the rays of the sun or some other light source.

Bar Code: Coded information printed on the film cassette that contains processing information for the automatic detection of the film manufacturer, the processing to be used, the film type, speed, and length.

Baryta Coating: Serves as a brightener layer immediately below the emulsion layer on photographic paper. Made from barium sulfate.

Bayonet Mount: A system of pins and channels used to position and lock a lens to a camera body.

Bellows: A light-tight, folding accordian-type sleeve that connects the camera body and the lens. Provides for considerable flexibility wherever used on photographic equipment.

Bleaching: Removal of image density from a developed and fixed photographic print.

Blue Sensitive Film: Those film emulsions that are sensitive to blue lightwaves and ultraviolet waves.

Bracket Exposures: Making one or more exposures on both sides of the exposure calculated to be correct. This helps to ensure success.

Bulk Loader: A device allowing the transfer of short lengths (20 and 36 exposure frame amounts) of film from a large roll such as 100 feet (30.5 m) to a reusable film cassette.

Bulk Loading: Using film from a large roll and placing it into reusable film cassettes in 20 and 36 exposure frame lengths.

Burning: See **Printing-In**.

Cable Release: A flexible cable used to trip the shutter on a camera without the photographer needing to touch the camera.

Camera Back: The portion of a camera that forms part of the body and is frequently used to help hold the film in position.

Camera Obscura: The observation of light rays passing through a small opening and forming an image on the wall of a darkened room.

Camera-Subject Axis: The imaginary line extending between a camera and the subject for exposure.

Camera Tilt: Refers to the view camera adjustments of the lens board and the film holder.

CDS: Initials meaning "cadmium sulphide," which when charged with electrical current is very sensitive to light. Used frequently in light meters.

Celestial Photography: Taking pictures of the moon, stars, sun, and other planets by using cameras attached to telescopes.

Center of Interest: The important part of a photograph that demands attention from the eyes of the observer.

C-41 Process: An abbreviation for "Kodak color negative film processing chemistry." It has become an industry standard that is used by many other companies.

Characteristic Curve: A graph used to describe the effects of light exposure or developing chemistry on the density of photographic film and paper.

Checker Code: Coded information printed on the film cassette. When loaded into a DX code equipped camera, it will automatically be read and the camera will be set for film speed, film length, and exposure range.

Chemical Concentrate: A chemical such as that used for photographic purposes that is much stronger than needed. In almost all cases, a photographic chemical concentrate is diluted with water to make a working solution.

Chemical Fog: A condition that occurs when the expected clear areas of a negative are unclear. This is due to the developer attracting the unexposed silver halides or when the fixer does not dissolve the unexposed silver halides.

Color Compensating Filter: Filters made of gelatin that are used to adjust selected colors during the process of exposing color prints in an enlarger.

Color Correcting Filter: Used to correct or change the color temperature of light to match the color balance of film in the camera.

Color Developer: A chemical solution that converts exposed silver halides to a dye in all three emulsion layers of reversal (slide) film.

Color Temperature: The level of radiant energy being emitted by a light source and measured on the Kelvin scale.

Color Wheel: A circular illustration showing the relation of the additive and subtractive colors to each other.

Commercial Processing Laboratory: A business that specializes in film developing, printmaking, and slide mounting.

Complementary Color: Opposite colors on the color wheel. A complementary color of an additive color is created by equally combining the other two additive colors.

Composition: Refers to a pleasing selection and arrangement of the elements within a photograph.

Computer-Aided Metering: An electronic system within a camera that automatically selects either or both the f-stop and shutter speed according to the light meter reading.

Concave (lens surface): The surface of a lens that is hallowed or rounded inward.

Condenser Enlarger: An enlarger that includes within the light chamber two or more bowl-shaped glass lenses that focus the light to the negative.

Contact Dermatitis: Inflammation of the skin due to contact with substances that cause the surface of the skin to react. Photographic chemistry can cause skin to itch, become red, develop small white sores, or other irritations.

Contact Print: Photographic prints made by placing a negative and print paper in contact and making the exposure. This makes a same-size print.

Contact Printer: A self-contained exposure unit used to expose same-size contact prints.

Continuous-Tone Film: Film capable of recording the many tones of gray from very light to very dark.

Converging Lenses: Lenses that cause refracted light to move toward a single point.

Converging Lines: Image lines that appear to converge or run together, such as how railroad tracks appear in the distance.

Convex (lens surface): The surface of a lens that is arched up or bulged out.

Copy Stand: A frame-like device that contains a base, lighting, a vertical post, and a bracket to securely hold a camera. Used primarily when taking close-up pictures of flat (two-dimensional) artwork.

Cropping: Removal of unwanted and distracting elements from the picture.

Dark Storage: A darkened area used by photography collectors and libraries to preserve photographic prints, negatives, and slides.

Dedicated Flash: A type of electronic flash that becomes fully integrated with the settings and action of the camera.

Depth-of-Field: The distance between the nearest and the furthest points that are in acceptable focus.

Developer: A chemical solution that reduces exposed silver halide crystals to black metallic silver. This causes the latent image in film and paper to become visible.

Dichroic Enlarger: A unit specifically designed for color printing that uses permanent glass filters and a mirror to project light.

Diffused Light: Light rays that have been scattered to produce an even distribution of light.

Diffusion: Scattering light waves while making an exposure with an enlarger. This softens the image lines, thus creating a pleasing print.

Diffusion Enlarger: An enlarger that includes within the light chamber a frosted piece of glass or plastic that diffuses the light before it reaches the negative.

DIN: The German film speed rating system. The initials stand for "Deutsche Industrie Norm." English translation—German Industrial Standard.

Direct Flash: Pointing an electronic flash unit mounted on a camera directly toward a subject when taking a picture.

Distortion: The undesired visual perspective of converged image lines when a camera is too close to a large object when the picture is taken.

Distortion Control: Refers to a view camera. Adjustment of the lens board and film holder allows the image shape to be completely controlled.

Diverging Lenses: Lenses that cause refracted light to spread to many points.

Dodging: Subtracting enlarger light during print exposure to one or more selected areas of the image. Purpose is to increase image detail.

Double Matting: Applying two window-like mounting boards over a photograph with one board being larger than the other.

Dry Instant Film: Processing chemicals are sealed within the layers of the film giving no visible signs of the chemical solution.

Dry Mounting Press: A unit having two platens that provide heat and pressure for attaching photographs to mounting board.

Dry Mounting Tissue: Thin, paper-like sheets having heat sensitive adhesive on both sides. Used to fasten photographs to mounting board.

DX: A code system printed on 35mm film cassettes that can be read by DX equipped cameras for automatically setting the camera for the film speed and film length. The coding is also used for automatic processing after the film has been exposed.

Easel: A device used to hold photographic paper in position while being exposed to light with an enlarger.

E-6 Process: An abbreviation for "Kodak Ektachrome Process E-6." A color reversal film processing procedure created by the Eastman Kodak Company that has become an industry standard.

EI: Letters used in abbreviation for exposure index. Used as a prefix to the number designating the exposure of film when pushed. At ISO 400, film pushed to 800 would be listed as EI800.

Electromagnetic Radiation: A series of electromagnetic waves within the electromagnetic spectrum.

Electromagnetic Spectrum: The entire range of wavelengths extending from gamma rays to alternating current electrical waves.

Electronic Flash: A gas-filled glass tube that when electrified produces a brief but brilliant flash of light. The light burst is the approximate color of midday sunlight. Sometimes, inaccurately called a strobe.

Electronic Imaging: The system of recording visual images on magnetic material through electronics.

Emulsion: The light-sensitive layer of photographic film and paper. It is a mixture of silver halide crystals suspended in gelatin.

Enlarger: A piece of equipment that is used to project images onto light-sensitive paper to make enlarged photographic prints.

Enlarger Head: The main part of the enlarger that includes the light chamber, negative carrier, bellows, and lens.

Environmental Portrait: A completely planned and posed portrait taken in the subject's business, office, or personal environment.

EVS: Initials meaning "exposure value system" and sometimes abbreviated to EV. The system uses a standard number for a selected combination of f-stop and shutter speed.

Existing Light: See **Available Light**.

Exposure Calculator: Gives suggested shutter and aperture settings based on a variety of scene descriptions and film speeds. Often made of two or more round dials fastened in the center so they will freely turn.

Exposure Guide Table: Provides scene or subject descriptions and the recommended shutter and aperture settings. Sometimes called an existing-light exposure table.

Exposure Meter: See **Light Meter**.

Exposure Meter Index: The scale (visual markers) within the camera viewfinder that is used to gauge the correct exposure.

Exposure Meter Needle: The small pointer used in an exposure (light) meter to identify the specific reading.

Extension Tube: Metal tubes of different lengths used to position a lens farther from the focal plane. This provides for an increase in the image size.

Eye Wash Station: A facility made up of a plastic bottle filled with a special eye wash, a mirror, and printed instructions. This should be located near the photographic darkroom area to be used for emergency chemical splash eye problems.

Face Mounting: Adhering photographs and other flat sheets to the surface of a single piece of mounting board.

Ferrotype: A chrome-plated steel plate or drum used to dry fiber-based photographic paper prints.

Fiber-Based: Refers to photographic paper in which the light-sensitive emulsion is applied to a base sheet made of cellulose (wood) fiber.

Fiber Optics: Very thin transparent fibers of glass or plastic that transmit light.

Fill Flash: Adding a limited amount (1/2, 1/4, or 1/16) of flash to light the shadow areas of an otherwise lighted scene.

Fill Light: The secondary light in a multiple light arrangement for portraits.

Film Advance: The system within a camera consisting of several parts needed to properly advance film before and after each exposure.

Film Cartridge: A light-tight container used for packing films used in cameras that use either 110 or 126 size film.

Film Cassette: A light-tight container used for packaging roll films such as the 35-mm size.

Film Clip: Small devices made of plastic or metal used to hang film for drying.

Film Contrast: The visual difference in density of the light and dark areas on developed photographic film and prints.

Film Developing: The processing stage that makes the latent image in film visible. One step in the series involving film processing.

Film Finishers: Chemical solutions used to reduce the washing and drying time of developed film. They also help to finish the developed film in a quality manner.

Film Grain: Clumps of developed silver halide crystals that form within the image area of photographic film.

Film Holder: The light-tight device used to hold sheet film in a view camera.

Film Leader: The preshaped lead end of a roll of film. One side of the film has been cut away for about 2″ (5.1 cm) from the end to make it easier to load in the camera.

Film Plane: See **Focal Plane**.

Film Pressure Plate: The spring-loaded flat piece on the inside back of a camera that holds the film securely against the film plane.

Film Processing: The complete series of steps necessary to make latent images visible and permanent on photographic film. The major steps include developing, stopping, fixing, and drying.

Film Protective Bag: Special film carrying bag containing a laminated lead coating to stop X-rays from exposing film at airport terminal passenger-check stations.

Film Retainer: The part in a camera that helps to hold a sheet or roll of film in the correct position.

Film Speed: The relative sensitivity of photographic film to light.

Film Speed Dial: The adjustable dial or wheel generally on top of a camera that is used to set the camera light-metering system to the correct film speed rating.

Film Sprockets: The rectangular holes on both edges of film such as 35 mm that are used to accurately advance the film through the camera.

Filter: A round or rectangular piece of material, usually glass or gelatin, that selectively absorbs and transmits light waves.

Filter Factor: A numerical system used to designate the needed increase in exposure due to the amount of light absorbed by a filter.

Filtration: Process of changing the color of light by using color printing filters in an enlarger.

First Developer: The first of two developers used in the Kodak Ektachrome Process E-6. Actually, it is a black-and-white developer that turns exposed silver halides to black metallic silver in reversal (slide) film.

Fish-Eye Lens: A special lens having a wide angle of view up to 180°. Also, it causes considerable image distortion.

Fixed-Focus: Relating to camera lenses that are preset during manufacture; thus, no operator adjustment is possible.

Fixer: A chemical solution made of sodium thiosulfate or ammonium thiosulfate, which dissolves the unexposed and undeveloped silver halides in film and paper.

Flare: Stray light that enters a lens and causes unwanted exposures on photographic film.

Flashbulb: Pre-electronic flash. These were airtight, clear or blue bulbs that gave a burst of light just as the camera lens shutter opened.

Flash Hot-Shoe: A bracket on top of a camera that holds the flash unit and contains electrical contact points.

Flash Meter: A hand-held light sensing device used to measure the burst of light from one or more electronic flash units.

Flash Synch-Socket: The electrical contact used to connect the flash unit wire cord to the electrical system of the camera.

Flat (lens surface):   The surface of a lens that is flat as compared to being concave or convex.

Flush Mounting: Adhering photographs and other flat sheets even with the edges of mounting board.

f-Number: See **f-Stop**.

Focal Length: The distance from the center of a lens to the focal plane when the lens is focused at infinity.

Focal Plane: The camera surface on which the focused image passing through the lens forms a sharp image. Often called the film plane.

Focal-Plane Shutter: One of two major photographic lens shutters that is made of two rubberized fabric or metal curtains. Primarily used in single-lens reflex cameras.

Focus-Assist Beam: Near-infrared or light beam emitted from a camera to aid the passive autofocus lens system.

Focusing Aid: A magnifying device used to critically focus an enlarger

Focusing Screen: The specially designed etched glass that permits the image to be seen in the viewfinder of a reflex camera.

Framing: The photographic composition guideline in which natural surroundings serve to focus attention on the main subject of the picture.

Freelance Photography: Skilled and experienced photographers who provide their services on a part-time basis or full-time self-controlled business.

Front Projection Backgrounds: A photographic method of producing backgrounds from in front of the subject when taking portraits.

f-Stop: A specific number setting or size adjustment of the iris diaphragm within a photographic lens. Also, a designation for a specific aperture opening.

f-Stop Numbers: Numbers representing fractions that relate to a mathematical ratio of the aperture to the focal length of a lens.

f-Stop Preselection: Selecting the f-stop for a specific purpose such as depth-of-field and then adjusting the shutter speed according to the light meter reading.

Gamma Rays: The shortest waves of energy within the electromagnetic spectrum. Not visible by the human eye.

Gelatin: A product made from cattle hides, hooves, and bones. When cool and dry, it becomes very hard. Used heavily in the manufacture of photographic film and paper.

GNP: Gross National Product. A measure of the total value of the goods and services produced by a nation in a specific period such as a year.

Graduate: A measuring container for liquid solutions. Preferably made of plastic; it has calibrated markings for ounces and milliliters.

Grain: The pattern produced by the silver halide crystals in photographic film. All silver-based film has grain. It is not visible unless the silver halides are large in the film emulsion and then increased in size again in an enlarged print.

Ground Glass: Glass having a frosted appearance created by sandblasting or chemical etching. Used in some camera viewfinder systems.

Hair Light: The small but concentrated light used to highlight all or a portion of the hair on the portrait subject.

Halation: The spreading of light beyond its planned or desired boundaries.

Half-Stop: Aperture adjustments midway between two standard f-stops on a photographic lens.

H & D Curve: See **Characteristic Curve**.

High Contrast Film: Film, either orthochromatic or panchromatic, that is formulated to record tones of gray as black or as white. Used primarily in the graphic arts industry.

Highlights: The lighter image areas of a photographic print.

High-Speed Photography: Photographs made of objects moving at a rapid speed such as a bullet shot from a gun.

Hot-Shoe: The bracket that is used to fasten an electronic flash unit on top of a camera. It contains electrical contacts which match with the flash, thus tying the electronic circuits of the camera and flash together.

Hydroquinone: One of the three chemical compounds commonly used as all or part of the reducing agent in developer.

Hyperfocal Distance: The minimum distance that is in focus when a lens is adjusted to the infinity focus position.

Image Deterioration: The reduction in quality of a photographic image due to light, heat, humidity, and other problems.

Image Points: The identified points that make up the image that is falling on the film focal plane.

Incident Light: Light from any source that is falling on a given area or subject.

Indicator Stop Bath: A special stop bath that contains a yellow dye. When the strength of the stop bath is depleted, the yellow dye turns a bright purple.

Indirect Flash: Deflecting or disbursing the light from an electronic flash unit so the light rays will not strike the subject in a direct fashion.

Infinity: Photographically relating to a camera lens adjustment that places all images in focus from a minimum distance to as far as the eye can see.

Infrared Film: Film that is sensitive to all visible lightwaves plus UV and infrared waves in the electromagnetic spectrum.

Instant Photography: Photographic equipment and products that give full imaging results in a few seconds to a few minutes.

Inverse Square Law: A law of physics that is used to describe how light intensity will be reduced by one fourth when a given distance is doubled.

Iris: The part of an eye that adjusts to control the amount of light that can enter. Likened to the aperture of a camera lens.

Iris Diaphragm: Overlapping thin metal leaves used to adjust the aperture opening of a lens.

ISO: The international film speed rating system. The initials stand for International Standards Organization.

Kelvin Scale: A system of measurement used to measure the color and quality of light. Zero degrees Kelvin (°K) is equal to -273.16 degrees centigrade (°C).

Key Light: The main and strongest light in a multiple light arrangement for portraits.

Latent Image: A hidden but known image that has been recorded in the emulsion of photographic film and paper.

Lateral Shift: The adjustment for moving the lens and film boards side-to-side on a view camera.

LCD: Letters that serve as an abbreviation for "liquid crystal display." Letters and numbers are formed in an information panel via computer chip technology on cameras, wrist watches, and other electronic products.

Leaf Shutter: One of two major photographic lens shutters that is made of three or more thin metal leaves that open and close to permit light to pass.

LED: Letters that serve as an abbreviation for "light emitting diode." For example, these are used to provide exposure data around the edges of the viewfinder in an SLR camera.

Lens: A piece of optical glass or plastic, ground or molded to a specific shape, designed for focusing images on the film (exposure) plane.

Lens Aberration: An optical defect in a lens, causing an imperfection in the refracted image.

Lens Element: A single piece of glass or plastic shaped to make it possible to focus light.

Lens Group: A series of single lens elements either cemented together or precisely positioned to better refract light.

Light-Color: Specific electromagnetic wavelengths that are distinguishable by the human eye and have been given names such as red, green, blue, etc.

Light-Direction: Refers to the line or course on which light waves are moving.

Light-Intensity: The saturation of light energy falling on a specified area.

Light Meter: An instrument designed to measure the amount of light falling on or reflected from a subject. The information is used to make the aperture and shutter settings on a camera.

Light Meter Acceptor: The "eye" of a camera metering system.

Light-Quality: Refers to the harshness or softness of the light rays.

Light-Tight: Does not permit light to enter. Refers to a container such as a film cassette or cartridge.

Loading Room: A small darkroom in which photographic film can be handled in total darkness.

Macro Flash: Using a macro lens, special electronic flash units, sensors, and equipment to take pictures of small objects.

Macrophotography: The process of using auxiliary and special lenses to take photographs of objects that are the same size (1:1) and up to 50 times larger than the original object.

Magnifier: A magnifying glass on TLR and some SLR cameras that helps in focusing the lens.

Manual Focusing Ring: That part of a camera lens that is used as a grip to turn the lens back-and-forth for precise focusing.

Manual Override: A useful feature on automatic and programmed cameras permitting manual selection of both the f-stop and shutter speed.

Mat Cutter: A device containing a sharp blade that is used to cut mounting board in various rectangular and circular shapes.

Matting: Covering the edges of a mounted photograph or flat artwork with a window-type frame cut from mounting board.

Medium Format: A camera that produces film images between the 35-mm and smallest view camera. Usually the image is $2\frac{1}{4}'' \times 2\frac{1}{4}''$ (6 × 6 cm) square.

Memo Holder: A bracket on the camera back where the film box lid can be held in place.

Metering System: The electronic light measuring system that is part of most adjustable and nonadjustable cameras.

Metol: One of the three chemical compounds commonly used as all or part of the reducing agent in developer.

Microprocessor: A small, but powerful, special-purpose computer.

Mode Dial: Adjusting knob that permits the camera operator to select the operating system for a camera.

Mounting: Adhering photographs and other flat sheets to special mounting board.

Mounting Board: A thick, hard paper product used as a base for adhering photographs and other flat sheets.

MQ Developer: A photographic developer that contains a combination of metol and hydroquinone as the reducing agent.

Multiple Element Lenses: A lens that is made up of several individual lens elements to form a single lens unit used in cameras and enlargers.

Nanometer: A unit of measure equal to one millionth of a millimeter.

Neck Strap: A sturdy, flexible strap made of cloth, leather, or plastic that is used to hang a camera from the photographer's neck.

Negative Carrier: The device, often made of two metal plates, that holds the negative in place during exposure.

Negative Color Film: Film that produces color negatives when processed. In turn, the negatives are used to make positive paper color prints. Sometimes called "color print film."

Negative Sleeve: Transparent or translucent plastic or paper fabricated to protect individual negatives and multiple-frame negative strips.

Neutral Density Filter: Absorbs light waves equally, thus does not change the color of light. Available in different density values.

Nodal Point: Considered the center of a lens element when viewing it from the side.

One-Hour Processing: Special self-contained and fully automated equipment permitting film and print processing in one hour. Generally associated with the amateur market.

Optical Disc: Light-sensitive discs used for high-density information storage.

Orthochromatic: Used to describe photographic film that is sensitive to blue and green lightwaves but not to red lightwaves.

Outdoor Portrait: A completely planned and posed portrait taken in an outdoor setting, such as in a park, along a river, or in a home backyard.

Oxidize: The combining of oxygen with a substance that reduces its strength.

Panchromatic: Used to describe photographic film that is sensitive to all visible lightwaves—blue, green, and red.

Panning: Keeping the camera in synchronization with a moving object while squeezing the shutter release to take a picture.

Parallax Error: The image difference from that seen in the viewfinder and that covered by the taking lens of a camera.

Passive AF: A camera autofocus system designed to determine the focus of the lens by analyzing the actual image scene.

Patterns: Repetitive designs created by nature and through objects or images prepared by humans.

Peel-Apart Instant Film: The original type of instant film that requires a separation of the print from the negative after about 1 minute.

Pentaprism: A five-sided prism that is an important part of a single-lens reflex camera viewing system.

Perspective Control: Using the four standard view camera adjustments to control image perspective.

Phenidone: One of the three chemical compounds commonly used as all or part of the reducing agent in developer.

Photo CD: An optical compact disc, when recorded with special equipment, that can be used to store up to 100, 35 mm photographs. The images (photographs) can be viewed on television and computer screens and color thermal prints can be made from them.

Photocell: An abbreviation for "photoelectric cell." A small device within a camera that reacts to light because of its electrical properties.

Photofinishing: The total process of developing film and making photographic prints by commercial processing laboratories.

Photogram: A photographic print made by placing objects on print paper and then exposing it to light. A film negative is not used.

Photographic Eye: Refers to a human having both learned and innate abilities to see a picture that will make a good photograph.

Photography: The art and process of recording images on light-sensitive film and paper.

Photojournalist: A photographer involved in news reporting.

Photomicrograph: A photograph of a magnified image of a small object or organism. Taken with the benefit of a microscope.

Photomicroscope: An instrument that combines a microscope, camera, and light source. Used to take photographs of very small objects and organisms.

Photo Minilab: A self-contained photographic processing laboratory used primarily for amateur materials. Often located in shopping centers.

Photoretoucher: A person skilled in correcting blemishes in photographic prints and film negatives.

Pinhole Camera: A camera of the most basic design in which a pinhole is used instead of a lens to focus the image onto the film

Pixel: A minute spot or space that makes up an electronic or photographic image.

Polarizing Filter: Designed to reduce reflections from nonmetallic surfaces and to provide contrast differences in cloud scenes.

Portfolio: A collection of work (photographs) arranged in a suitable order and contained in a large folder or case.

Portrait: A photograph of people, animals, or birds taken in a posed formal or informal setting.

Positive Color Film: Film that produces right-reading color images directly in the film. Sometimes referred to as reversal film because of the developing. Used to create direct-viewing slides. Also, called transparency film.

PQ Developer: A photographic developer that contains a combination of phenidone and hydroquinone as the reducing agent.

Preservative: One of the four subparts of a developer solution. Its purpose is to keep the reducing agent and the accelerator from spoiling.

Pressure-Sensitive Mounting Sheet: Adhesive material in sheet form used to adhere photographs and other flat artwork to mounting board.

Printing Frame: A wood or metal frame containing glass that is used to hold negatives and photographic paper in tight contact for making exposures.

Printing-in: Adding enlarger light during print exposure to one or more selected areas of the image in order to increase image detail.

Print Proofer: A hinged glass frame designed to hold negatives and print paper in firm contact when making contact prints and proofs.

Print Tongs: "Finger-like" devices used to transfer photographic prints from one processing tray to another.

Processing Kit: A container, usually a cardboard box that includes all needed chemical concentrates to process color film.

Processing Room: A darkroom outfitted with the necessary equipment and tools needed to process photographic film.

Processing Tank: A round or rectangular-shaped, water-tight container used to develop photographic film. Roll film tanks are light-tight.

Product Photography: Photographs made of products that are used for information and promotion through advertising literature, magazines, and books.

Professional Photography: Accomplished by highly skilled and experienced photographers who make their living via photography.

Programmed Camera: A camera containing a microcomputer that because of its programming during manufacture will select both the f-stop and shutter speed according to the light meter reading.

Pupil: The portion of an eye that focuses the reflected light into a sharp image. Likened to the lens of a camera.

Pushing Film: Exposing photographic film at a speed higher than its ISO rating.

Rangefinder: A focusing and distance system built into some cameras; commonly used in adjustable viewfinder cameras.

Rear Nodal Point: The theoretical center of a multiple element lens.

Rear Projection Backgrounds: A photographic method of producing backgrounds from behind a special rear projection screen when taking portraits.

Reciprocity Failure: The breakdown of the reciprocity law when the exposures are extremely short or very long.

Reciprocity Law: The balancing relationship between exposure length and aperture opening when various combinations equal in theory the same amount of light.

Reducing Agent: One of the four subparts of a developer solution. Its purpose is to turn the exposed silver halides to metallic silver.

Reflected Light: Light from any source that is reflected from a given area or subject.

Reflex Viewer: An attachment that fits on a view camera that allows the image to be seen without using a focusing cloth.

Refraction: The bending of light waves as they pass from one medium such as air into another medium such as the glass of a photographic lens.

Relative Humidity: The ratio of water vapor present in air to that possible at a given temperature.

Remote Flash Sensor: A device that senses a bright flash of light from one flash unit and triggers one or more electronic flash units.

Replenishing Developer: A developer that can be brought back to full strength after using it to develop a roll or several sheets of film. A special replenisher solution is used.

Resin-Coated: Refers to photographic paper in which the cellulose (wood) fiber base is made resistant to liquids by the application of transparent plastic coatings.

Restrainer: One of the four subparts of a developer solution. Its purpose is to keep the reducing agent from attacking the unexposed silver halides.

Reticulation: The separation of the emulsion from the film base; caused by a sudden temperature change in processing chemistry and water.

Retina: The portion of the eye that records the images and causes vision to occur. Likened to film used in a camera.

Retouching: Correcting problems in a negative or print; caused by dust and dirt or carelessness while making the print.

Retractable Flash: A flash unit, usually small in size, that is part of the camera housing. It can be lifted for use and retracted or returned to its in-camera location after use.

Reversal Bath: A chemical sensitizer used to convert light-sensitive silver halide crystals to a latent image in reversal (slide) films. The latent image eventually becomes the colorful images seen in slides.

Reversal Film: Color film that produces a positive image after processing. See **Slide Film**.

Rewind Button: Small button in the base of an adjustable camera that releases the take-up spool so that film can be rewound into the film magazine.

Rise and Fall: The adjustment for raising and lowering the film and lens boards on a view camera.

Selenium: A nonmetallic element used in basic light meters because it reacts to light intensity.

Sensing Cell: A unit on an electronic flash that measures the amount of light reflected from the subject. It turns the flash off when the correct amount of light has been emitted.

Shadow: The absence or reduction of light in a given space; caused by an opaque object blocking light rays.

Sheet Film: Photographic film that has been cut into individual pieces. Typical sizes include 4 × 5, 5 × 7, and 8 × 10 inches.

Shutter: A mechanism within a camera that admits measured amounts of light needed to correctly expose film.

Shutter Preselection: Selecting the shutter speed for a specific purpose, such as to "stop action," and adjusting the f-stop according to the light meter reading.

Shutter Priority: A camera feature that allows the photographer to manually set the shutter speed and then the camera-computer selects the appropriate aperture.

Silhouette Photograph: A photograph made with the subject between the camera and light source. This causes only the outline shape of the subject to be visible.

Silver Bromide: An extremely sensitive compound used in the preparation of photographic emulsions.

Silver Chloride: A mildly light-sensitive compound used in the preparation of photographic emulsions.

Silver Chlorobromide: A combination compound of chloride and bromide used in the preparation of photographic emulsions.

Silver Halides: The light-sensitive particles or crystals that make up a photographic film and paper emulsion. A compound containing silver mixed with bromine, chlorine, or iodine.

Single-Lens Reflex Camera: A major camera category; contains a single lens for both viewing the image and taking the picture.

Skylight Filter: Has a pink tinge of color that helps give better skin tones for outdoor portraits.

Slide Film: Color film that produces a positive image; often called reversal film or transparency film. Most frequently made into 2 × 2 inch (5 × 5 cm) slides.

Slide Mounter: A device or machine used to secure individual exposures of processed reversal (slide) film in 2 × 2 inch (5 × 5 cm) cardboard, plastic, or glass frames.

SLR: Initials meaning "single-lens reflex" camera.

Snapshot Photography: Photographs that are taken quickly without preplanning; also, economical cameras requiring little photographic skills are often used.

Snapshots: Pictures taken quickly without consideration for composition and lighting.

Sonar Focusing: The use of sound waves to focus the lens of a camera.

SPD: Initials meaning "silicon photodiode," which when charged with electrical current is very sensitive to light; used frequently in light meters.

Special Effects Filters: Filters specifically designed to create unnatural or unavailable images and colors in selected scenes.

Spectrogram: A photograph or a line diagram of all or part of the electromagnetic spectrum.

Specular Light: Strong, harsh waves of light that cause shadows with sharp edges.

Spindle: The center shaft around which film is spooled in a 35-mm film magazine or in any other type and size of film packaging.

Split-Image Focusing: A method of focusing using split halves of a circle etched into a camera focusing screen.

Spot Meter: A light meter designed to measure reflective light from a very narrow angle of view.

Spotting: Removing small imperfections in a photographic print that have been caused by dust and dirt specks on equipment and film.

Squeegee: A tool containing one or more rubber blades that is used to remove excess water from photographic film and prints.

Stabilization: A special process that requires only a two-chemical bath for processing photographic prints.

Stabilizer: Generally used as the final solution with color reversal and color negative film processing. Contains compounds that improve the stability of the color dyes and help to keep water spots from forming on the film while drying.

Stabilizer Solution: An acid-based solution used to neutralize the activator solution in the stabilization process.

Stop Action: Using fast film and a fast shutter speed to stop the motion of a moving object while taking a picture.

Stop Bath: An acid-based solution that is used to neutralize the action of the alkaline base developer solution.

Streak Camera: Extremely high-speed cameras that use electron beams to gather and project the fast-moving images.

Strobe: A misnomer used to identify electronic flash. See **Electronic Flash.**

Studio: A facility designed and equipped to permit the photographer's control over lighting and environmental conditions.

Studio Portrait: A completely planned and posed portrait taken in the controlled environment of an indoor studio.

Studio Props: Devices such as chairs, tables, wagon wheels, and toys that are useful when taking portraits.

Subtractive Colors: The secondary colors of yellow, cyan, and magenta. Each is produced by equally combining two additive colors. Combining all three subtractive colors produces near black.

Sunlight: Light emitted from Earth's largest star—the sun.

Supplementary Lens: A separate lens that can be added to a regular camera lens giving enlarged images for close-up work.

Swing Control: The adjustment for turning the film and lens board left to right on a view camera.

Tacking Iron: A small electrically heated tool used to adhere dry mounting tissue in small areas.

Taking Lens: The lens of a TLR camera that gathers the image and focuses it on the image plane.

Teleconverter: A special-purpose lens used together with the regular camera lens to give extra image magnification.

TLR: Initials meaning "twin-lens reflex" camera.

Tonal Values: The measured density areas of a photographic print, ranging from the lightest highlights to the darkest shadows.

Tone: Visual appearance of the surface color of photographic paper.

Toning: Changing the color of a black-and-white print to one of many colors—sepia (light brown), green, red, or blue—using bleach and pigment solutions.

Transmit: To allow or cause something to go through something else, such as light passing through a filter.

Trap Focus: An autofocus lens feature which releases the shutter when the subject enters the preset focusing distance. Sometimes referred to as freeze focus.

Tripod: A three-legged, adjustable camera stand.

Tripod Socket: The threaded opening in the base of a camera, allowing it to be secured to a tripod with a special bolt.

TTL: Initials meaning, "through-the-lens."

Twin-Lens Reflex Camera: A major camera category; contains two sets of lenses—one for taking pictures and the other for viewing the image.

Ultraviolet Filter: A colorless filter that absorbs ultraviolet light waves.

Ultraviolet Focusing: The use of ultraviolet light waves to focus the lens of a camera.

Unipod: A single-legged, adjustable camera stand.

Universal Language: A means of communication that can be understood by people of different nations. Photography is such an example.

Variable Contrast: Refers to photographic paper in which image tonal densities can be controlled with the use of special filters in an enlarger.

View Camera: A major camera category; it is a large format camera that permits viewing and focusing of the image on a ground glass.

Viewfinder: The camera part that permits an operator to view the desired picture-taking subject prior to releasing the shutter.

Viewfinder Camera: A major camera category; it is generally considered a camera for the amateur photographer.

Viewfinder Eyepiece: The opening for the eye to look through the camera viewfinder.

Viewing Glass: The ground glass upon which the image is formed in view cameras and reflex cameras without a pentaprism.

Viewing Lens: A camera lens system that permits the operator to see the image about to be photographed.

Viewing System Hood: Refers to reflex cameras that have a pop-up lid covering the viewing screen.

Vignetting: Feathering image edges in a photographic print to create an image that stands out strong, but fades gradually to nothing.

Visible Grain: The clumps of developed silver halide crystals that can be seen on enlarged photographic prints by the naked eye.

Visible Spectrum: That portion of the electromagnetic spectrum between 400 nm and 700 nm that can be seen by the human eye.

Visual Communications: The group of communications methods and processes in which information is transmitted by graphic means. Photography is an important aspect of visual communications.

Visual Perspective: The condition of seeing a scene, object, or building in its natural manner and trying to capture that same sight with a camera.

Wave: The curving pattern of the flow of energy in the electromagnetic spectrum.

Wavelength: The measurable distance between one point and the next corresponding point in the direction of travel of an "energy" wave.

Weight: Relates to the thickness of photographic paper: light weight, single weight, medium weight, and double weight.

Window Matting: See **Matting**.

Working Solution: A chemical solution, such as that used in photography, that is ready to be used. In some cases as with film developer, a working solution can even be diluted with water before it is used.

Wratten: The name of a system used to identify the kind, color, and density of photographic filters.

X-Ray Exposure: Nonvisible light waves that expose photographic film.

X-Ray Photograph: A shadow picture made with X-rays on special film; used frequently in medical science.

X-Rays: Short waves within the electromagnetic spectrum that are used in exposing special sensitive photographic film.

Zoom Lens: A lens with several moveable elements that makes it capable of many focal lengths within its limitations.

FIGURE CREDITS

The following companies, institutions, and individuals provided the identified photographs and line drawings.

SOURCE	FIGURES
AGFA Corporation	24-11
American Airlines	2-15, Chapter 1—Header
Armed Forces Institute of Pathology	1-6, 3-1
Bel-Art Products	25-5
Bencher, Inc.	36-9
Berg Color-Tone, Inc.	41-2
Berkey Marketing Companies	2-14
(Charles) Beseler Company	31-1, 36-10, 39-5, 39-10
Bestwell Optical Instrument Corporation	31-4
Byers Photo Equipment Company	37-8
California Stainless Mfg.	27-9
Calumet Photographic, Inc.	11-2, 11-6
Canon U.S.A. Inc.	9-7
Celestron International	3-10, 3-11, 5-3
Chinon America, Inc.	8-3
Clark's Photo Art Studio	22-12
Colenta America Corporation	5-6, 39-4
David W. Coulter	18-14, Chapter 4—Header

Creatron Inc.	41-11
Barton A. Dennis	13-11, 23-1
E. A. Dennis	37-1, 39-1
(The) Denny Mfg. Co. Inc.	22-3, 22-7
Dial-A-Photo System	23-4, 23-5, 23-6
Dimco Gray Company	27-2, 31-3
Doran Enterprises, Inc.	27-8, 27-11, 31-2, 31-6, 31-8, 31-10, 31-11, 31-14, 33-8, 34-3, 40-1
Marla Dory	18-2
Eastman Kodak Company	1-2, 1-7, 1-8, 2-13, 3-7, 4-2, 4-3, 4-11, 5-1, 5-8, 5-11, 5-12, 5-15, 6-3, 6-4, 6-13, 6-14, 8-2, 12-8, 19-12, 21-2, 23-3, 23-12, 24-6, 24-10, 24-12, 28-10, 29-1, 30-4, 30-6, 30-7, 33-12, 38-1, 39-14, 44-3, 44-4, 44-6, 44-13, 44-14, 44-15, 44-16, Chapter 8—Header
Edric Imports, Inc.	12-3
Electronic Photography and Publishing, Sony Business and Professional Group	44-5, 44-7, 44-8, 44-9
ESECO Speedmaster	31-18
Anthony F. Esposito, Jr.	11-10, 11-11, 11-12
Sharon K. Fahey	18-1
Falcon Safety Products, Inc.	6-10, 6-11, 12-5, 27-7, 33-10, 40-2, 40-3, 40-5, 40-13
GMI Photographic	13-17
René C. Gallet	1-5, 18-7, 26-5
Julie Habel	36-1
Hadland Photonics, Ltd.	44-10, 44-11, 44-12
(Victor) Hasselblad	9-15
Heico Chemicals, Inc.	26-4
(Karl) Heitz, Inc.	5-4
(The) Holson Company	2-1, 41-7, 41-8, 41-9
Hope Industries, Inc.	37-6, 37-7, 38-5
Ilford, Inc.	24-2, 24-8
Instrumentation Marketing Corporation	3-8, 3-9, 5-5
Alan Jackson	12-2
Jobo Fototechnic, Inc.	27-4, 27-5, 27-12, 31-13, 35-13, 37-4, 38-3, 39-8, 39-9, 39-11, 39-12, 39-13, 39-16

Richard L. Johnson	23-9
JVC Company of America	44-1, 44-2
J. R. Karsnitz, *Graphic Arts Technology*	36-4
Birdie Kramer	22-13
Kelvin K. Kramer	17-9, 18-15, 18-17, 23-2, 23-8
Kinetronics Corporation	31-15
Kleer Vu Plastics Corporation	42-5
Kreonite, Inc.	27-1, 39-18, 41-3
Labex Engineering Corporation	39-7
Leedal Inc.	22-14
Leica Camera, Inc.	8-9
(E.) Leitz, Inc.	3-3, 3-4
Light Impressions Corporation	29-7
Lisco Products Company	11-8
Louisiana State University Medical Center, Department of Pathology	3-2
Lowel-Light Mfg. Inc.	21-17, 21-18
LTM Corporation of America	21-14
3M	36-8
Mamiya America Corporation	6-7, 6-8, 8-7, 10-1, 11-1, 11-9, 17-4
Michael Business Machines, Corp.	40-4
Minolta Corporation	9-2, 9-11, 13-4, 13-8, 16-15, 16-16, 17-5, 21-6
(The) Morris Company	1-9, 20-14, 23-13
MWB Industries, Inc.	31-19
Mystic Color Lab	4-7, 4-8, 4-9, 4-10, 5-7
National Aeronautics and Space Administration	Chapter 7—Header
Nikon, Inc.	1-11, 6-5, 6-9, 9-1, 9-5, 16-8
Noritsu America Corporation	4-12
Novatron of Dallas, Inc.	21-15
Olympus Corporation	1-10, 2-3, 8-4, 9-16, 16-2, 16-3, 16-4, 36-11
Omega Arkay	11-7, 31-5, 31-7, 31-9, 32-1, 34-2
Optische Werke G. Rodenstock	4-13
Pentax Corporation	2-4, 6-6, 9-8, 9-9, 9-12, 9-13, 9-14, 12-6, 13-13, 13-14, 13-15, 13-16, 15-4, 21-3, 21-8
Photo Control Corporation	5-2, 22-6, 22-10, 39-17
(The) Pinhole Camera Company	7-6, 7-7, 7-8, 7-9, 7-10, 7-12, 36-12

INDEX